PENGUIN BOOKS

## Ice Rivers

Jemma Wadham is Professor of Glaciology at UiT the Arctic University of Norway, the Norwegian Polar Institute and the University of Bristol. She has led more than twenty-five expeditions to glaciers around the world, including to Greenland, Antarctica, Svalbard, Chilean Patagonia, the Peruvian Andes and the Himalaya, and has won several prestigious national awards for her research, including a Philip Leverhulme Prize and a Royal Society Wolfson Merit Award. She is best known as a pioneer in the field of understanding glacier-hosted life and the impacts of glaciers on our global carbon cycle. Ice Rivers is the first book she has written for a general readership.

# Ice Rivers

*A Story of Glaciers, Wilderness and Humanity*

JEMMA WADHAM

PENGUIN BOOKS

PENGUIN BOOKS

UK | USA | Canada | Ireland | Australia
India | New Zealand | South Africa

Penguin Books is part of the Penguin Random House group of companies
whose addresses can be found at global.penguinrandomhouse.com.

First published in Great Britain by Allen Lane 2021
First published in Penguin Books 2022
001

Printed and bound in Great Britain by Clays Ltd, Elcograf S.p.A.

The authorized representative in the EEA is Penguin Random House Ireland,
Morrison Chambers, 32 Nassau Street, Dublin D02 YH68

A CIP catalogue record for this book is available from the British Library

ISBN: 978–0–141–99414–7

www.greenpenguin.co.uk

Penguin Random House is committed to a
sustainable future for our business, our readers
and our planet. This book is made from Forest
Stewardship Council® certified paper.

# Contents

**Leverett Glacier, Greenland**
- 600 km²
- Land-terminating outlet glacier/polythermal

**Shallap and Pastoruri Glaciers, Cordillera Blanca, Peru**
- 7 km²/<5 km²
- Valley glacier/temperate

**Steffen Glacier, Patagonia, Chile**
- 420 km²
- Lake-terminating valley glacier/temperate

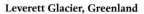

**Finsterwalderbreen, Svalbard**
• 34 km$^2$
• Valley glacier/polythermal

**Haut Glacier d'Arolla, Swiss Alps**
• <3.5 km$^2$
• Valley glacier/temperate

**Chhota Shigri, Indian Himalaya**
• 16 km$^2$
• Valley glacier/temperate

**Joyce Glacier, Antarctica**
• 40 km$^2$
• Valley glacier/cold-based

# Introduction

## Icy Beginnings

Glacier, n. 'A slowly moving river or mass of ice formed by accumulation of snow on higher ground'

*Oxford English Dictionary*

Imagine that one morning you woke up, pottered into your kitchen to make a cup of tea, found you hadn't shut the freezer properly last night, and ice was now teasingly protruding out of a crack in the door. And that the next day the ice had grown, bursting open the freezer door, and starting to advance across the floor, then over the counters, effortlessly sweeping up the toaster, kettle and dirty dishes into its icy folds. Then a day later it had engorged the entire kitchen and started creeping as a giant, dripping, frozen tongue, upstairs. A week later it had filled the entire house, its icy fingers pointing like antennae skywards through the fractured window frames, then continuing its merciless advance down the street, soon to entomb your city, your country, your continent. Now imagine the minute amounts of meltwater produced at the edges of this vast body of ice, pooling together to feed tumultuous rivers the size of the Nile, finally to disgorge their watery load into the Earth's seas, shaping what life thrives, how ocean currents flow and, sometimes, whether our climate warms or cools. This is not myth – this is the scale of glaciers, almost beyond what a human mind can grasp.

My interest in glaciers was first awakened while roaming the Cairngorms in Scotland as a teenager. I was intrigued by the bald, ashen grey hills, sculpted by the passage of glaciers at the height of Earth's last intense cold period, nearly 20,000 years ago. The valleys were unusually broad – scoured by the ice, presenting lumpy terrain at their edges where glaciers had eroded, unceremoniously dumped and then modelled soft sediments into clutches of giant eggs, called 'hummocky moraine'. The thought of a snake-like torso of moving ice hundreds of metres thick having once upon a time advanced down these valleys astounded me.

Yet my fascination with the Cairngorms was not coincidental. Over the first fifteen or so years of my life, their weather-ravaged slopes had become a source of freedom for me, where structures were erased and I could feel the pulse of something much, much bigger than myself. I would trudge with purpose to the summit of my favourite peak, Ben Gulabin at the Spittal of Glen Shee, barely able to draw in enough air to fuel my leaden legs. Austere steely-grey crags loomed from the toe of the hill softened by the rusty tones of its heathery scalp – this panorama swept me away from what had become a disconnected and sometimes bewildering upbringing.

The confusion found its roots, I suppose, after I lost my father in a car accident on Christmas Day when I was eight. Children in those days were kept away from funerals, we never talked about the event much afterwards, and I felt numb for many years, aware only of a strange sense of disconnection. One moment I had a father, the next he was gone without trace. I created my own world in my head, immersing myself first in novels, in a fantasyland of characters and places, and then later in the Cairngorms, where I came to find serenity and calm – just 'me' and 'the mountain'. The barren sweeping landscapes gifted me connection to 'something bigger', as I struggled to find my footing in an increasingly tumultuous family life.

Yet against this backdrop, the kernels of my future as an explorer and glaciologist also found their beginning. By the age of eleven, I was organizing family holidays, writing away to hotels and holiday cottages enquiring about booking. '*Dear Mr Woodman*,' they wrote back . . . always misspelling my name and utterly oblivious to the fact that I was but a girl still at primary school. Our first grand excursion was to the Lake District – I had recently learnt to sail a dinghy and wanted to test my skills – including my brother, who was recovering from an emergency appendix operation. I persuaded my mother to hire a sailing boat from which we ventured jauntily out onto the ripples of Derwent Water, until we came to ground on the mudbanks, narrowly avoiding capsizing. My brother was huddled in the bows clutching his side in pain as the boat veered over, and my mother, in her holiday finery, was left wading through murky waters up to her waist. Even at this young age, an independent instinct had started to develop, and I began to nurture a desire to go 'out there' and discover Earth's great wilderness.

From then on, mountains grounded me and helped me breathe – and drew me in, too, like a story in a book I never wanted to end. First in the Cairngorms, then through GCSE Geography classes, where I learnt about the huge rivers of ice that had advanced down my yawning Scottish glens, and then finally to university. I spent many hours delving deep into the curricula of every Bachelor's Geography degree in the UK, determined to pinpoint the one with the highest 'ice content'. Cambridge came top of my list, and by some apparent miracle I managed to do well enough to get there. This led me to the Swiss Alps, where I first laid eyes on a real glacier as a twenty-year-old university student – glaciers in the flesh were beyond anything I had imagined. White, pristine and unpolluted hinterlands, a blank canvas, capable of absorbing any negative emotion that pulsed through me and miraculously transforming it into pure exhilaration and joy.

Since then, I've followed that very same smell of the ice and its vertiginous white expanse. I've grown to know and understand glaciers better, and with that depth of learning, as always, comes heightened fascination, perhaps even obsession. In 2012, after nearly twenty years of toil in the field, I became a professor (of glaciers) at the age of thirty-nine. Yet, it has often felt that my life and my journey with glaciers have woven their way like two rambling paths tracked across a mountain – we come together, we exchange words, we part for a while, only to return again. These twisting threads have led me all around the world and back again, my goal to piece together clues to help understand how glaciers behave and what meaning they hold for us as humans. A grand detective story.

The fascinating thing about glacier ice is that it is not quite like the clear ice cube in your gin and tonic. 'Glacier Blue' has become almost a cliché in the paint-chart world – and yet it's not always blue. It can be blue or turquoise, certainly, when emerging from the depths under great pressure at glacier margins, after years of slow compaction have squeezed the air bubbles from its body, rendering it blue because the one colour the ice does not absorb well is indeed – blue. If you think of light as a rainbow of colours flooding through the skies, different objects on the Earth's surface have different abilities to soak up these rays of coloured light (or 'energy', we could say) – the rays that they don't absorb well are reflected and give the object its colour. So forests reflect green, glacier ice with few air bubbles reflects blue, and snow reflects everything and so is white (i.e. colourless). However, glacier ice can also appear bright white if it contains a lot of air, or dirty brown when speckled with sediment picked up from its rocky underlay – here it becomes less 'glacier' and more something else.

Peer deeper into the ice, under a microscope, and you may be surprised to see not a dull, vacuous mass, but many elongated hexagonal crystals formed from hundreds of water molecules standing side by side like soldiers, their borders defined by tiny water-filled microscopic channels ('veins') kept from freezing by the very high concentrations of salts dissolved within them. Apply pressure to glacier ice and its crystals deform and dislocate, their watery veins acting as slip planes and allowing the glacier to flow. This property – flow – is what differentiates a glacier from an ice cube. The ice deforms under its own immense weight and, river-like, the glacier slowly slips down over its mountainous terrain.

But this is just the start – a myriad of streams dissect glacier surfaces and descend via cavernous vertical shafts ('moulins') to feed deep subterranean rivers which emerge explosively through ice caves at the glacier edge, their violent flows tumbling ever downwards until they meet the ocean. Upon first glance, glaciers seem so silent, passive and lifeless; and yet, measured over decades, centuries and millennia, they are some of the most sensitive and dynamic parts of our planet, growing during ice ages and shrinking under the malign influence of our carbon-choked atmosphere. Their cyclic growth and decay over the last two million years, in response to very subtle shifts in the way the Earth orbits the sun, has caused our sea levels to fall or rise by over one hundred metres, as vast amounts of meltwater have been stored by, or released from, ice sheets blanketing North America, Europe and the Antarctic – enough water to drown the Statue of Liberty.

Towards the end of 2018, I was rushed to hospital with a benign brain cyst the size of a tangerine. I had recently secured my first big job, as director of a research institute. My life was a chaotic blur of meetings and events, my friends described me as 'manic', and I felt utterly exhausted as I struggled to make a success of my new position – it was a far cry from my serene icy wildernesses and glaciers. I wouldn't allow myself time off to visit a doctor, despite the fact that I was experiencing explosive headaches, starting to lose my sight, had numbness in my legs and couldn't walk in a straight line down a corridor. This may sound a little mad, and to this day I'm not sure what stopped me from investigating further. I expect it had something to do with fear (most things do) – fear of failing at my job, of letting people down if I took time off, and perhaps even fear that behind my strange symptoms lurked something rather serious. Then, BANG – I was suddenly in A&E, and within twelve hours I was out cold on an operating table, my skull cut open in an attempt to rid my brain of the enormous, life-threatening growth. Over the months that followed and as I recovered, I grappled to comprehend what had happened to me, and to re-evaluate what I truly cared about. This led me straight back to my old friends, the glaciers.

You see, our glaciers are not far off where I was in December 2018. They are amid an acute health crisis of their own, melting at unprecedented rates, as our climate warms year on year. Fossilized carbon (oil, gas, coal) takes millions of years to form, as layer upon layer of dead plants and animals are slowly bedded down and stored in the deep, but we have been returning this ancient carbon to the atmosphere in just a matter of decades. Rising greenhouse gases like carbon dioxide have already warmed the Earth by one degree Celsius since we started burning fossil carbon at the beginning of the 'industrial' era (about 150–200 years ago);[1] more terrifying still, we are on course to hit a colossal three degrees Celsius or more of globally averaged warming by the end of this century.[2]

Already the impacts of this warming are being keenly felt on glaciers. In 2019 record melt rates were reported on the Greenland Ice Sheet; Himalayan glaciers were found to be thinning at much higher rates than scientists had thought; and the first obituary was written for a glacier in Iceland. I've witnessed this accelerating melt – glaciers I have studied in the European Alps have shrunk back more than a kilometre since I first visited them twenty-five years ago. It's easy for me to believe in climate change, but it's less easy for those who haven't seen the drip, drip, drip of the ever-lengthening summer melt and the vast lakes which are forming precariously in the wake of the retreating ice – pinned only by rubble and the moving glacier, these lakes are quick to burst their banks if they become overfull with meltwater. Our news feeds are bombarded daily with reports of waning glaciers, but I can appreciate that to many people these are impersonal tales with little meaning. *Oh, there goes another glacier, how sad!* Do we ever act purposefully to save something if we experience it only through dry facts and figures, but lack a connection to it in our hearts?

None of us knows how long we have left on the planet – I thought I had decades left until that moment at the end of 2018 when my life seemed like it could be over in a flash. Nor do we know how long our glaciers have left – but certainly most of the glaciers in the European Alps will be gone by the end of this century if we continue to burn fossil fuels at the current rate. So my intention is to introduce you to the glaciers, to share the emotional connection that I have fostered with them during three decades of research. You see, to me, glaciers are not just moving bodies of ice. Each one has a unique character deriving from the way it flows, melts and is framed by its incredible wilderness. When I'm with them, I feel like I'm among friends. My return to them in this round-the-world voyage heralds a return to my old self. A kind of personal re-wilding – borders dug up, earth left untilled, seeds of ideas allowed to drift freely in on the wind and to take root to sprout new, vibrant green shoots. A story of glaciers and people, their histories and mine, entwined. It is, in many ways, a love story.

# PART ONE

# The Smell of the Ice

# 1. Glimpses of an Underworld

*The Swiss Alps*

I was twenty, on my first expedition, a somewhat green Geography undergraduate working as a field assistant on a research project that aimed to uncover mysterious details about the flow and plumbing of the Haut Glacier d'Arolla, a small, relatively accessible valley glacier tucked up high in the Swiss Alps. I had pored for hours over the theories of glaciers in geography textbooks, of course, and was familiar with their handiwork from family holidays in the Cairngorms – but it was here that I would meet one for the very first time.

I had come completely unprepared, with a small rucksack full of mostly summer clothes, my brother's old army boots (several sizes too big) and a plastic mac which had served me well in Scotland but boasted the breathability of a crisp packet. Camped at 2,500 metres above sea level in the rocky valley of the Haut Glacier d'Arolla, I had spent my first night sleeping on cardboard in an old sleeping bag I'd used for sleep-overs when I was eleven, with its thin walls of clumpy polyester fibres and all the heat retention of a hessian sack. At this point I'd never heard of Polartec, or Gore-Tex, or even the concept of a Karrimat. Constantly disturbed by the muffled roar of the glacial river not far below and the shotgun-like cracks of rockfalls on the slopes above, not to mention the thin air which laboured my breath and the cold that made my bones throb with pain, I'd barely slept. Suddenly I understood why glaciers had been considered the resting place of ghouls and evil spirits in medieval times.

So this is where it all began – my journey as a glaciologist. Of course, I was following the deeply grooved path of many before me. The European Alps have always been a prime stomping ground for glaciologists, with glaciers of all shapes and sizes mostly accessible by foot – from the elongated, streamlined twenty-kilometre-long ice tongue of the Swiss Aletsch Glacier to tiny, stubby glaciers which are barely notice-able, perched high up in concave rock depressions (cirques) above the wide plains far below. Spanning 1,000 kilometres between Nice in the west and Vienna in the east, the Alps are part of a much greater mountain system, the Alpides, which stretches as far as the western Himalaya. Mountains are always a sign of geological drama, and so it is for the Alps, which formed as the African plate began to creep north into the European plate around 100 million years ago.

During their most intense collision, around thirty million years ago, the two plates squashed old crystalline basement rocks and younger seafloor sediments from a pre-Mediterranean ocean, neatly folding them into a series of vertically stacked 'nappes' – rather like the sail of a boat when hauled in to be stored on the boom, fold overlapping fold. The rock was crum-pled most vigorously in the western Alps, where the mountain belt is thinner but higher, and includes such giants as Mont Blanc – at 4,800 metres the pinnacle of western Europe. During the past two million years, the Alps have been reshaped and remoulded by intense phases of glacial erosion as the Earth has plunged in and out of long cold periods (glacials) and short warm periods (interglacials), which reflect natural oscillations of our climate caused by tiny shifts in the Earth's orbit of the sun.

It was Jean-Pierre Perraudin, a mountaineer and hunter from

Lourtier in the Valais region of Switzerland, not far from the Haut Glacier d'Arolla, who posited one of the first modern theories of glaciation.[1] He speculated around 1815 that oddly smoothed rock surfaces were caused by glaciers effectively 'sanding' the rock they flowed over, with any protruding rocks and stones in their basal ice layers gouging deep grooves in the direction of the ice flow. He observed that giant boulders strewn across the valleys near his home were of a foreign rock type, and must have been dumped there by a glacier when ice filled the valleys during the last glacial period. Although Perraudin had an intimate understanding of the mountains, still he had to toil against the prevailing belief of the day, which was that great biblical floods had been the protagonists in forming the alpine landscape. It seemed inconceivable to him that a flood could have dislodged and transported these giant boulders, which would clearly sink like stones. He spoke to the naturalist Jean de Charpentier about his findings, but de Charpentier dismissed them as 'extravagant' and 'not worth considering'.[2]

It took another fourteen years for Perraudin's theories about glaciers to be fully developed, first by Ignace Venetz, a highway and bridge engineer in Val de Bagnes and another native to the Valais region of Switzerland. He had attempted to create channels to drain meltwaters from a large lake which had grown at the edge of a local glacier when its ice advanced and dammed a stream – such glacier advances were common during such times and were a symptom of the final throes of a cold snap during the Middle Ages in Europe, popularly called 'the Little Ice Age'. However, Venetz failed in his attempts, and the lake catastrophically flooded the valley and destroyed many lives and houses.

Venetz had many conversations with Perraudin about the inner workings of glaciers. By 1829 he was finally convinced, and presented his ideas at the annual meeting of the Swiss Society of Natural Sciences, which argued that the glaciers of his time were all that remained of a much larger mass of ice that once covered the Alps. This time, Jean de Charpentier supported him, now also swayed by these theories of massive glaciation. Yet it was Louis Agassiz, a Swiss biologist and geologist who grew up near Fribourg and ended up as a Professor of Natural History at the University of Neuchâtel, who, through a mixture of serendipity and determination, brought the early theories of glaciers to the fore in his famous *Études sur les Glaciers* in 1840. Agassiz is often lauded as the grandfather of glaciology, but in truth there were several, starting with Perraudin. They all applied pressure to the wall of conventional wisdom, until the wall weakened and ultimately collapsed.

The first time you wake up somewhere new in the mountains is always the most explosive for the senses. Dragging myself out from beneath my humble canvas on that first alpine morning, I was greeted by a panorama that remains one of the most memorable of my life. Directly across the valley an imposing mass of ice tumbled over a col (the saddle between two peaks) and down the seemingly vertical rock wall about five hundred metres in height – not a waterfall but an icefall, where the glacier meets the end of its hanging valley and must venture over the precipice below.

Here the glacier in question, the Bas Glacier d'Arolla, flows quickly down over the steep rock face, stretching until its tiny crystals can no longer deform fast enough to permit flow as a single mass, and the ice fractures in a million planes to form a chaotic field of crevasses and sharp ice towers, known as seracs. Icefalls are death traps to the mountaineer. Perhaps the most notorious example of an icefall can be found in the upper reaches of the Khumbu Glacier, the highest glacier on Earth, which moves at about a metre per day. This is one of the most treacherous parts of the ascent from base camp to Mount Everest's summit in the Himalaya. Climbers can take as long as a day to pick their way through the glacier's tortuous path, and it has caused several dozen deaths over the last fifty years – simply because ice flows, and the faster it flows the more difficult it becomes for it to move as a single body, leading to crevasse fields and icefalls.

A rather incredible feature of glaciers is that they have been found to flow in three possible ways, the slowest of which is through the deformation of glacier ice crystals. Ice behaves more like a liquid than a solid; technically speaking, ice is a 'viscous fluid', or a 'non-Newtonian fluid', which means that its viscosity (or gloopiness) depends on its temperature and how much pressure it is under; the greater the pressure and the warmer the ice, the gloopier it becomes, and the more its crystals squash or 'deform'. Glaciers grow ever-deeper over time as snow accumulates, and compression plus a little melting and refreezing turn it to ice, after which the crystals start to deform under pressure. By this means, a typical alpine mountain glacier like the Haut Glacier d'Arolla might move just a few metres per year.

All glaciers flow by means of the imperceptible deformation of ice crystals, but they have much quicker ways of moving too. A second means by which glaciers flow involves the glacier sliding over a wet, slippery rock surface. Imagine taking an ice cube from the freezer, placing it on a flat plate, then tilting the plate – the ice cube slides off, right? Now consider the same ice cube on the same tilted surface, but still in the freezer – it's going nowhere, because the cube is frozen to the plate and there's no liquid water to lubricate its flow. Small glaciers in the Arctic, where the air is very cold, behave like the frozen ice cube – they don't slip and slide. They can only move by their ice crystals deforming. But in warmer climes, like the Alps, where the bottoms of glaciers have a thin layer of water, they can slip over their beds – these are called 'temperate glaciers'.

Even in perishingly cold places like Antarctica, very thick glaciers can curiously still have liquid water at their beds. Imagine blowing up our ice cube to monstrous proportions, the size of a skyscraper hundreds of metres in height, but still keeping it in a freezer. (It's a very large freezer!) Will it move if you tilt the surface? Actually, it might. Remember that old physics experiment where you hold a cheese wire across a block of ice, then apply pressure to the ice through the wire? The pressure lowers the melting point of the ice, and the wire slices down through it. Thus, our gigantic ice cube – a bit like the Antarctic Ice Sheet – will probably melt at its base due to the huge pressure of the overlying ice. Then, if by some superhuman feat you manage to tilt the surface upon which it rests, it will start to slide – in the wonderful world of glaciology, this is called basal sliding.

A glacier has a third ingenious way to flow if it rests on top of wet mud (or sediment, as a glaciologist would probably call it). Imagine that we now slid a tray of very wet soil collected from the garden just after a heavy rainstorm beneath our vast ice cube in the outsize freezer. What happens next? Well, the pressure of all that ice pressing down on the wet soil causes the water in its tiny pore spaces to become pressurized, which lowers the friction between the soil grains. This makes the soil weak and easy to deform, so if you tilt the tray, the soil will move like a mudslide downhill. The ice cube rides majestically on top of this moving platform of wet, deforming mud. This mechanism of ice flow is known as sediment deformation.

So, ice deforms, ice slides, sediments beneath the glacier

deform – that's three ways glaciers can flow. Putting it in human terms, some glaciers crawl (ice deformation only), other glaciers walk (ice deformation and basal sliding), and a few virtually sprint as ice deforms and the glacier slides, perhaps also hitching a ride on top of deforming sediments. Small glaciers in the European Alps can be considered walking glaciers. In an average year the Haut Glacier d'Arolla moves at most ten or so metres in its centre, where the ice is not slowed down by dragging against the rock sidewalls.[3] However, the speed of the ice can more than double for brief periods in summer when meltwater crashes to the base of the glacier, pumping at such high pressure that it pushes up against the ice and raises it slightly off its bed, a process called 'hydraulic jacking'.[4] What's common to all glaciers with water at their beds is that the processes controlling much of the glacier's flow mostly happen in this inhospitable abyss known as the subglacial zone.

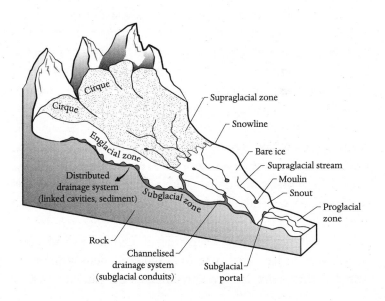

The grand purpose of the expedition I was part of was literally to get to the bottom of the Haut Glacier d'Arolla. This compact valley glacier – so described because it neatly inhabits its valley without overflowing it – formed the focus of the Cambridge University-led 'Arolla Project', headed up by one of my all-time glaciology heroes, Professor Martin Sharp, with the audacious goal of discovering what lies beneath glaciers by simply *going there*. In the case of Arolla, this meant somehow passing through one hundred metres of solid, moving ice.

For me, one of the most enthralling things about glaciers is the fact that the place where all the action happens you can neither see nor touch. You are left to imagine the point where the ice ends and the rock begins, and ponder what life could survive such grinding hostility as the glacier moves, picks up and regurgitates boulders, stones and sand. Only when the ice retreats is the evidence revealed, an ornate assemblage of ice-etched, polished rock surfaces, carved melt channels, moulded sediments – traces of a past dark, violent underworld.

You sense a glacier long before you set foot on it – it makes itself known in the sharpness of the air. But first, reaching the front of any glacier (commonly known as its snout or terminus) normally involves a monotonous hike through what is known as the 'proglacial zone'. Here, a mass of boulders, pebbles, sands and silt present a chaotic scene, parading sediments and rocks regorged by the glacier during cycles of advance and then retreat, like a person who has upped sticks in a hurry, leaving their house in a mess. Only the ribbons of milky rivers and jewels of emerald lakes trapped between moraines offer signs of life in this barren, uneven moonscape. You barely notice much beyond your feet, while dodging holes and other unsavoury features such as sinking mud, which is commonly found in the vicinity of river channels – fine glacially eroded sediments masquerading as waterlogged leg-sucking death traps. If you prod the surface with a pole, it wobbles like a blubbery belly – 'porker mud', we called it, a term you won't find in the glossary of any textbook.

Meanwhile an icy wind increasingly penetrates your lungs; if anything can be both exhilarating and foreboding, this is it. That first tantalizing smell of the ice, that sense of being stroked by its soft, frigid fingers, is a welcome and a warning. This 'katabatic' wind (from the Greek word *katabasis*, meaning 'descending') often builds through the day; it results from heavier, ice-cooled air flowing down the glacier to its snout. Such winds are often seen by mountain communities as the spirit of the glacier.[5] For me, they are a sign to prepare, to put on an extra layer and get ready for a laborious climb up the steep front of the glacier.

If I'm honest, my first sighting of the Haut Glacier d'Arolla was something of an anticlimax. In fact, I wasn't even sure that it was a glacier, for it was barely distinguishable from its rocky surrounds. Glacier snouts are grubby things, 'snout' being an appropriate word here, given that it essentially has its nose buried in mud, rather like a foraging pig. As glaciers move, they pick up sediments and stones from their rocky beds and receive rockfall from their surrounding valley walls, transporting all this debris, then releasing it upon melting in their lower reaches. Since melt is always highest at the snout of a glacier simply because it's warmer at lower altitude, the release of this grey debris from its icy shroud is fastest here, and consequently mounds of dirt accumulate chaotically at the margins and fronts of glaciers in ridges known as moraines. Glacier snouts might seem motionless and silent, almost dead, but still the ice continues to flow, albeit slowly, only advancing if the glacier as a whole receives more snowfall in a year than it loses through icemelt.

And so I tentatively set foot upon my very first glacier. The front of a glacier is treacherous terrain, often full of holes and cracks caused by the collapse of the roofs of channels which convey meltwater beneath the thinning terminal ice. As far as snouts go, that of the Haut Glacier d'Arolla was not particularly steep – high rates of melting and relatively slow ice flow maintained a gently sloping profile. But for someone whose exercise regime involved gentle pedalling around the flats of Cambridgeshire, it still came as something of a shock. I began to climb slowly, following the steady footfall of my more experienced companions up the eastern middle ('medial') moraine, which guaranteed me safe passage onto the bare ice. Such moraines are common on alpine glaciers, formed by rockfall off the head and side walls of glacier valleys onto the ice. The fallen rock becomes buried by successive winter snowfalls in the upper reaches of the glacier; the fragments are transported along with the ice until they re-emerge on the glacier's lower flanks, where melting brings them once more to the surface to pile up in elevated ridges, protecting the ice below from melt. When seeking to walk up onto a glacier, you're well advised to find yourself a moraine.

Eventually the steepness of the snout abated, my breathing slowed, and my jellified legs regained their equilibrium. Gingerly, I stepped off the moraine and onto the ice. The first steps on ice signify an important moment of union for me. Regardless of how many glaciers I have set foot on, the feeling is always reliably vivid. The repetitive crunch as the brittle surface crust shatters beneath your feet, combined with the mesmerizing sensation of walking upon a block of shifting ice some hundreds of metres thick – the sense of mystery and danger never dulls.

The meandering, ice-walled streams on the glacier surface,

which beam cool turquoise in the sun, might seem benign enough – until they end abruptly and flush their watery load into deep vertical shafts, moulins, which lead straight to the bed. Staring into the abyss of a moulin always gives me vertigo – I try to stop myself, but it's almost impossible not to imagine 'what if' scenarios. What if I just slipped and plummeted head first down this vertical hole? What if, what if, what if . . . *stop*! The key thing about moulins, though, is that by passing vast volumes of water to the dark depths, they provide the only teleconnection between the glacier surface and the bed.

After an hour or so of trudge and crunch, a curious spectacle appeared on the horizon – a collection of silhouetted figures, beetling about around a humming machine which, on closer inspection, was itself framed by an untamed, writhing umbilical mass of black hose, giant water tanks and assorted tools. One figure carried a vertical lance-like pole, a jet of water and steam spewing explosively from its end. This was the famous drill site – where a bunch of resourceful scientists from the University of Cambridge had solved the challenge of accessing the icy depths of the Arolla Glacier using an extremely long hosepipe. The hot water was emitted as a high-pressure jet from a metal drill nozzle directed down vertically on to the ice, stuttering to life with the vigour of a car-wash powerhose. Gradually, over some hours, it bored a hole about as wide as a tea plate through hundreds of metres of glacier ice to its deep, murky underbelly.

Until the early 1990s, there had been scant attempts to reach glacier beds. The pioneering Arolla Project aimed to perforate a small rectangle of the glacier surface (a few hundred metres by fifty metres) with holes that stretched from surface to bed, and then to drop pressure and temperature sensors down these holes to gauge how water flowed beneath the ice. A typical borehole was almost invisible when walking over the ice surface, its location only revealed by the jumble of wires protruding out of it, tethered to various devices at the glacier bed. I would gaze down these holes, mesmerized by the swirl of cool blue and white glassy ice walls getting darker and darker until they disappeared to black nothingness. The thirty or so boreholes that were drilled to the bed of the Haut Glacier every year told us so much about how glaciers worked. Over time, the continual sliding and deformation of the ice around the boreholes caused them to stretch from vertical to more of a banana shape, since the ice at the top of the column was flowing faster than the layers below it. This is due to the cumulative effect of all the different glacier flow types, which add up as you move from the glacier's bottom to its surface.

But I was always most fascinated by the story that boreholes told me about water. Most boreholes when you peered down them appeared strangely dry; yet, while being drilled, they could be brimming with water. As soon as the drill nozzle hit the glacier bed, though, the water would often mysteriously vanish, sometimes slowly, sometimes rapidly as though someone had pulled out the plug. This disappearance of water indicated that the hole had intercepted some form of active drainage system at the glacier bed, such as a river channel (or 'subglacial conduit', as a glaciologist might call it), with ice walls and a rocky base.

The faster the water drained and the lower the water level in the borehole plummeted post-drilling, the more likely it was that the drill had intersected a channel that could rapidly transport the water down-glacier. These channels were capable of gulping more and more meltwater; they were able to grow by simply melting back their ice walls, which meant they could keep their meltwaters flowing at low pressures, leading to low borehole water levels. That said, sometimes things happened too quickly for conduits to do this. Vast fluctuations in surface melting and water flow to the glacier bed between day and night caused the water levels of the boreholes that pierced these channels to swing wildly by as much as one hundred metres (almost as thick as the ice) over just one day.[6] Bordering these super-efficient channels at the glacier bed were swamp-like zones of tiny interconnected waterways, which only allowed the sluggish flow of meltwater, on a mission to eventually arrive at a fast-flowing channel. Technically called a 'distributed drainage system' (see figure, p. 9), they coped well with a small, constant drip feed of meltwater, but tended to over-pressurize, then melt out when flooded with meltwater. When this happened, super-efficient fast-flowing rivers or channels formed in their place.

Boreholes had the pitfall of only supplying information about what was going on at a single point, rather than the entire glacier bed – which, in the case of the Haut Glacier d'Arolla, covered several square kilometres. But there were other techniques which provided a more 'zoomed out' picture of what was happening beneath all that ice. Dye tracing was one of them. Early on in my Arolla days I crossed paths with Pete Nienow, a tall, lanky Cambridge PhD student who was legendary for his ability to jog up and down the steep mountain path between the Arolla village and our camp in under forty minutes, in espadrilles, barely breaking a sweat. Pete's job was to build this 'big picture' view of the glacier plumbing system, which he did using a harmless bright pink tracer dye called rhodamine.

Each day Pete would beetle up the glacier, a tiny bag of the powdered dye stashed in his rucksack, until he sighted a moulin. He would then swiftly sprinkle the pink stuff into the water gushing down the moulin, before sprinting back down the glacier to eagerly await the dye's arrival in the main river bubbling out from the glacier's snout – by which point the dye had been so diluted that it no longer turned the water pink, but could still be detected by a fluorometer. This instrument, by shining a small beam of light into the water, then measuring the amount of light that was absorbed and re-emitted by any pink particles, could compute how much dye there was in the river. The faster the dye appeared and then disappeared in the river at the glacier front, the quicker and more efficient a flow path it must have taken to get there. Pete did this all summer for several years, identifying some thirty moulins dotted over the glacier surface.

These dye-tracing experiments revealed that, over the course of a summer, as the snowline retreated up the glacier and melt rates soared, the slow, rambling passage of melt through sluggish passageways at the glacier bed was soon replaced by rapid flow through efficient channels[7] – in effect the glacier bed's network of rural tracks collapsed and was traded for giant motorways. Pete's results from across the entire glacier complemented the experiments with 'single-point' boreholes, which taught us via their water levels how the speedy and sluggish parts of the glacier plumbing system interacted on a daily basis. During the day, ice on the glacier surface melted, supplying water to super-efficient conduits at the bed, which overflowed, forcing water into the channel margins and the inefficient drainage networks beyond; at night the flow reversed, returning water to the channels.[8]

It was an ingenious plumbing system, but by no means perfect. In spring, when the first pulse of fresh snowmelt gushed to the bed via moulins and crevasses, the internal pipework couldn't accommodate this sudden influx of water, and the force of all the pressurized water literally jacked the glacier off its bed. Imagine the curious spectacle (if only we could see it with the naked eye) of the entire glacier surface lifting, rising on a pillow of meltwater, and, as it lifted, suddenly the glacier loosening its contact with its rocky bed, removing the friction that kept it from racing downhill, and surging forward.[9] These 'spring events' happen on glaciers the world over, but are soon curtailed by the formation of channels at the glacier bed which whisk away the ponded meltwater, allowing the glacier to nestle back onto its rocky base and slow down.

In the evenings, after long days working on the Haut Glacier d'Arolla, during the slow plod down over the ice and through the glacier forefield, I always felt a sense of deep connection with the barren mountainscape, as the fatigue softened my senses. As I descended, to the east, turning a soft pink in the sunset, stood the jagged Dents de Bouquetins – a classic arête, or knife-edged ridge, towering as much as one thousand metres above the glacier below, its sharp features derived from powerful erosion when much of the valley was occupied by the glacier at the height of the last 'cold' period, about 20,000 years ago.

The ridge is named after the bouquetin, the enigmatic alpine ibex (*Capra ibex*), which looks like a deer but is more closely related to a goat. Bouquetins are native to much of the European Alps and traverse dizzying heights, usually close to the snowline, their split hooves enabling them to clamp tenaciously onto the underlying rock. Occasionally, at dusk, I'd glimpse the distinct horned profile of one poised on the cliffs above. These graceful mountain goats display what is called sexual dimorphism – a fancy way of saying that the males and females look different. The males can have spectacular horns which curve backwards, continuing to grow throughout their lives,[10] sometimes up to a metre in length. The size of a male bouquetin's horn determines its place in the hierarchy. The females are smaller, with less dramatic horns. The sexes mostly live apart, reuniting during the mating season in late autumn. The horned goat features in the star sign Capricorn, which happens to coincide with the main breeding period for the bouquetin in December, straddling the winter solstice. Could people in ancient times have associated their mid-winter celebration of fertility and rebirth with the breeding cycles of this sure-footed mountain dweller?[11]

These elusive animals have long beguiled alpine communities; their body parts were believed in medieval times to boast magic, medicinal powers, and they were also once prized for their meat. Indeed, the stomach remains of the 5,000-year-old iceman 'Ötzi', who was found preserved within a glacier high in the Ötztal Alps in Austria in 1991, contained remnants of cooked ibex meat.[12] The goats also feature in prehistoric cave paintings in France, such as the Chauvet-Pont-d'Arc cave in the Ardèche from 30,000 years ago. However, following the invention of firearms in the fifteenth century, the tale of the bouquetins was a sad one – they were hunted to close to extinction. They only thrive today due to conservation measures and campaigns to re-introduce them.

Bouquetins can be confused with their smaller cousins, the chamois, which have much smaller, curved tipped horns and partially white faces. Chamois haunt the lower mountain flanks – I often spotted them during my numerous hikes up the path from the Arolla village to camp. I would always hear them before I saw them. First, there was a clatter of rock fall, and looking up, if I was lucky, I would catch a glimpse of one of these bold, nimble creatures athletically leaping across some near-vertical precipice, seemingly defying gravity.

That first summer at Arolla was a heady time of my life.

I remember practising ice-axe arrests by throwing myself down a practically vertical slope; learning how to rope up to cross crevasses; and, most essentially, teaching myself how to imbibe throat-scorching *eau de vie* at altitude while still commanding control of my legs. These were high-octane experiences. The communal mirth of field-camp life was infectious, I had never laughed as much before – in short, I was hooked.

Even so, it was demanding. Day after day, we followed the same drill. Get up, thaw out over a brew, trough down a bowl of gritty muesli (the sand produced by glaciers defies all barriers, literally penetrating everything), toil through the moraines, then clamber up onto the ice. Often the build-up of heat during sunny days was so intense that electric storms rumbled through the night – shocks of lightning and thunder reverberating around the cavernous valley, echoing off the steep rock walls, booming and repeating as if on an interminable loop. I recall these epic storms as black and white negatives etched in my mind; the glimpse for a millisecond of the tumbling icefall illuminated by flashing light, like a sinister beast caught prowling in the darkness. The next day, I'd wake and it was as though nothing had happened – just a bad dream and ghosts of memories in monochrome.

Our small camp consisted of a humble collection of sun-bleached canvas tents, nestled in a hollow in the mountainside overlooking the Bas Glacier d'Arolla icefall. It was encircled by barren rocky slopes, adorned with clumps of rough grass that were grazed by herds of scruffy sheep which wandered in and out without warning. Often in the middle of the night you would wake startled by their scuffling and snuffling beneath your tent fly in search of tasty morsels. The camp was situated just above the glacial river, its raging torrent always audible as white noise. Its flow was lowest first thing in the morning, after a cold night of low melt on the glacier, reaching a climax in late afternoon – but day or night, it continued to whoosh its fine milky suspension of sediments downstream, eventually feeding into the Rhône, then Lake Geneva and onwards, finally discharging its load into the Mediterranean in the south of France near Arles.

Many of our great rivers originate as tiny trickles of snow and icemelt in the high mountains – which begs the question, what will happen once the ice that feeds them has all melted? It's a question of special importance in regions where glaciers are melting and human communities are numerous and vulnerable, such as the Himalaya and the Andes, for the pulverized rock in these rivers, known as 'glacial flour', has been shown to be a fertile source of nutrients. The farmers of the Swiss valleys give credence to this notion by habitually spraying the glacial meltwater onto their croplands to ensure a bountiful harvest. But the one thing you must never do is drink the milky liquid. The very fine rock powder which gives glacial rivers their cloudiness can in some glaciers contain harmful levels of heavy metals like arsenic, mercury and lead, and some of its minerals can easily irritate the stomach lining.

The sheer power of the river emerging from the Haut Glacier d'Arolla became apparent to me early on during that first summer. I was standing on the bank one day, preparing to cross from one side to the other, holding a prism on a stick as a colleague surveyed its position from the river terraces above. (A survey prism reflects beams of light sent out by a surveying instrument – in this case a geodimeter – back to the meter, which is normally sited somewhere high up and with a wide view of what you want to survey. The time taken for the signal to arrive back is used to measure the distance.) I started to shuffle across, moving ever so slowly so as not to lose my footing, cautiously placing the prism in position lest I dropped it, and finally reaching the central main flow. The force of the wall of water pushing against my spindly rubber-coated legs was immense. Suddenly I lost my footing and found myself bouncing downstream over the cobbled riverbed, tossed by the turbulence of the ice-cold water, heading directly for the intake point for the hydro-electric power station. I could barely think, let alone panic. I tried rolling into an upright position, but this was challenging in the chaotic flow. Eventually I was lucky enough to be fished out by another team member, a tall, ginger-haired chap called Mark, who spotted my blue and white dry suit bobbing down the river. Thus my first glacier fieldwork romance was sparked.

Bidding farewell to the Haut Glacier d'Arolla at the end of that first summer expedition was, for various reasons, an emotional experience – I keenly felt the loss of the close companions who, only weeks earlier, had been complete strangers. A few years after my departure, an astounding discovery emerged from the Arolla Project. Martin Sharp (now based in Canada), working with UK microbiologists and chemists, had found life at the bottom of boreholes drilled to the bed of the glacier, stunning the global research community.[13] There are sometimes discoveries that you simply don't anticipate, but when you look back, you can't comprehend why not. It's obvious – why wouldn't you expect life beneath glaciers? There's plenty of water, which is a prerequisite for life. But still, it came as a bolt from the blue, because the mindset of the glaciologists (mostly physicists and geographers) had been shattered through collaborating with a bunch of biologists. Such revelations demonstrate the value of working together across the traditional boundaries between disciplines, because it's at these boundaries that ideas are catalysed.

Okay, so in the case of the Arolla glacier, we are not talking about the discovery of anything greater than can be spotted down the lens of a microscope – but still, here were microorganisms, the most adaptable and resilient life on Earth. As much as the published article was lauded across the world, those of us who knew the glacier recalled that the borehole site wasn't far down from the pipe responsible for evacuating human waste from the Refuge des Bouquetins on the precipice above. However, my expeditions since have shown that, actually, wherever you look in glaciers, you do in fact find life – surviving against the odds, and using every clever biological trick in the book to do so. How does this life survive and function, and what impact does it have beyond the glacier? These are mysteries that I have spent the last twenty years trying to solve.

In 2018, twenty-six years after my first encounter as an enthusiastic but somewhat green undergraduate, I revisited the Haut Glacier d'Arolla. I trudged the steep path from the Arolla village at 2,000 metres above sea level to the spot where our camp had once stood, eagerly anticipating the sight of my beloved glacier. The climb was easier than I remembered, probably reflecting my higher level of fitness, having long since traded the flats of Cambridgeshire for the rolling hills of the West Country. Returning to the place where my passion for ice had first taken hold was both exhilarating and moving. The memories flooded back – of life amid our tiny cluster of sun-scorched tents, the daily toil up the glacier, even the position of particular boulders – but one thing truly stunned me. The ice-fall, which I recalled as a colossal flowing mass spanning the mountain from top to bottom, was barely recognizable. Its tongue, which once tumbled as a single crevassed torso down the steep rockface, was now drastically severed in the middle, such that its upper part was no longer connected to the lower tongue of the Bas Glacier d'Arolla in the valley below. Thus, the Bas Glacier d'Arolla, which the icefall had previously nourished, was about one kilometre shorter. I had witnessed the death of a glacier.

Up and over the lip of the hanging valley, in the main valley once carved by the Haut Glacier d'Arolla, I was again dumbfounded – where was the glacier? The landscape was painted in drab greys and browns, as if in mourning, apart from high white splotches where seasonal snowfall had yet to melt. Squinting, I could just make out the tiny brown snout of the Haut Glacier. Compared with twenty-five years ago, it was a kilometre further up the valley, and it sat motionless like a dark ghost, the steep, rocky sidewalls forming a shroud around its lower flanks. I was aghast – it was as if I'd returned home, only to find it had been ransacked. My stomach was in knots, and tears of disbelief welled up in my eyes.

Climate change might at times feel like an abstract concept, but when you experience a spectacle like this, and can compare your own before-and-after photos, the conclusion is undeniable. Such observations are echoed in the scientific literature, for satellite images reveal more than 20 per cent loss of glacier area in the Swiss Alps between the early 1970s and 2010.[14] But how do we know this to be the fault of humans? Couldn't it just be caused by natural cycles? Certainly, the planet has seen dramatic swings in temperatures, which have caused glaciers to both grow and shrink. During the Cenozoic Era, spanning the past sixty-five million years, over which our continents drifted to roughly their present positions, the climate (bar a few blips) has cooled, causing the gradual growth of glaciers. A key player in these shifts has been the concentration of carbon dioxide in the atmosphere, and changes in what we call 'the greenhouse effect'. This effect describes the fact that when the sun's rays hit the land surface they warm it up; some of that energy is radiated back to space, only not all of it makes it back, and instead is absorbed by clouds and greenhouse gases (one of which is carbon dioxide). This energy, or heat, is effectively trapped, as in a greenhouse.

The amount of carbon dioxide in the atmosphere over very long time periods such as the Cenozoic – before humans got to work, that is – has been a tug of war between several countervailing forces.[15] Volcanos add carbon dioxide to the atmosphere, erupting as the Earth's tectonic plates shift around. The breakdown of carbon-rich rocks at the Earth's surface also adds a little. Meanwhile, carbon dioxide is removed from the atmosphere by the weathering of other rocks and the growth of plants. Plants on land and plant-like creatures in the ocean are direct consumers of carbon dioxide during photosynthesis, and if their dead remains become buried below ground or in the depths of oceans – in the form of carbonate rocks such as limestone or perhaps as undecayed organic matter in peatlands and permafrost – this burial removes the gas from the atmosphere for a long period of time.

These different push-pull forces are thought to interact – for example, if volcanos spew out more carbon dioxide, this stimulates more weathering and plant growth which consume the gas and helps stop the planet getting too warm. It's a bit like a thermostat. Over the last four hundred million years, as life moved out of the oceans and on to land, the expansion of land plants and the associated weathering of rocks are thought to have gradually taken more and more carbon dioxide out of the air, helping prevent the Earth from heating up too much as a result of volcanic carbon dioxide.

During the Cenozoic a melting pot of different forces and

their impacts on atmospheric carbon dioxide have slowly nudged Earth into the most recent of its so-called great 'Ice Ages' – times with significant ice at the poles. At its beginning, around sixty-five million years ago, continents were in their final phase of wandering away from their mother super-continent, Pangaea, hitching a ride on top of tectonic plates. India at this time was detached from Eurasia, and the Indian plate dived down beneath the European plate by a process called subduction. High volcanic activity at the point where the plates met heated up carbon-rich rocks, producing carbon dioxide that flooded the atmosphere – more than a thousand molecules per million of carbon dioxide,[16] compared to just over four hundred per million today.[17] Earth's climate was hot – too hot for the build-up of ice.

But starting from around fifty million years ago, India ran into Europe, pushing up the Himalaya mountain belt and the Tibetan Plateau – there was no more plunging of one plate beneath the other, and the carbon dioxide emitted by volcanos gradually fell. There's some debate as to whether the weathering 'sink' for carbon dioxide grew or not. It's possible it was boosted by the expansion of plants and some large-scale movement of plates; possibly the uplift of the Himalaya caused high rates of erosion, as the mountains were worn down by the elements and glaciers, and attacked by carbon dioxide in rain and meltwater.[18] Whatever the cause, carbon dioxide concentrations have fallen since about fifty million years ago and Earth has shifted from a 'greenhouse' climate to an 'icehouse' one.

Around thirty million years ago, the climate was cool enough to form a large ice sheet in Antarctica.[19] This was aided by the creep of Australia, South America and Africa away from the Antarctic continent, creating a sea way and a powerful ocean current which swirled anti-clockwise (the Circum-Antarctic

Current) around its land mass, keeping Antarctica cool. Two to three million years ago, following further cooling and carbon dioxide decline, ice sheets grew in the northern hemisphere, starting with the Greenland Ice Sheet that still remains today. Once ice sheets grow to a large size, their white, twinkling surfaces help keep the climate cool by reflecting as much as 90 per cent of the sun's rays back into space.

The last two million years of the Cenozoic are known as the Quaternary Period. During this time carbon dioxide in the atmosphere was at an all-time low in the Cenozoic and conditions were already cool. Our records show that a new factor became important in the climate's tug of war at this time, producing a fresh suite of climate variations on top of long-term Cenozoic cooling. Tiny, regular shifts in the shape of the Earth's orbit of the sun began to influence Earth's glaciers, by affecting the amount of heat reaching the surface of the planet, and acting as the pacemaker for regular cycles of warming and cooling.[20] These orbital variations were not new, but started to have more impact once the ice sheets were large and the climate was cool. There is also some evidence that they became more pronounced around three million years ago, causing cooler northern hemisphere summers which allowed ice to build up.[21] Thereafter, they led to a pattern of swapping between short warm 'interglacial periods' of tens of thousands of years to longer cold 'glacial periods' of up to 100,000 years. Glaciers and ice sheets grew during glacial periods – including ice sheets across North America and Europe – and then melted again during interglacial periods, in cycles called 'glacial-interglacial cycles'.

We are currently sitting in an interglacial of the Quaternary period called the Holocene, and have been for roughly the past 10,000 years. There have been naturally occurring climate variations during this period – we know, for example, that during 'the Little Ice Age' of the Middle Ages, temperatures were as much as two degrees Celsius lower than those found today. Many glaciers grew during this time, continuing to do so until the mid-nineteenth century. Countless paintings depict glaciers advancing down their valleys, engulfing entire alpine villages; bishops were even summoned to exorcise the evil spirits from these ice monsters.[22]

So climate change has clearly occurred naturally in the past, both warming and cooling. However, the alarming truth is that atmospheric levels of carbon dioxide and other greenhouse gases such as methane have soared over the last century. We know this from the tiny air bubbles trapped in ancient ice laid down in the middle of Antarctica and Greenland, sampled by drilling deep cores down to layers that go back almost a million years, covering as many as eight glacial-interglacial cycles.[23] This is largely due to human activity – burning fossil fuels, cultivating rice paddies, felling forests, rearing livestock, to name but a few. Some people have argued that we can't call the current period the Holocene anymore, since humans have created their own climatically distinct epoch – the Anthropocene.

The present-day concentrations of carbon dioxide in the atmosphere are now as high as during the mid-Pliocene, three million years ago;[24] global mean air temperatures were up to three degrees warmer than today and sea levels twenty metres higher – as if the Greenland and West Antarctic Ice Sheets largely disappeared, plus some ice from around East Antarctica.[25] Which begs the question, where are we headed?

As we pump more and more greenhouse gases into the atmosphere, we have to delve ever further back in time to find a period where similarly greenhouse gas-charged conditions were present on Earth – only then it was due to the rising influence of volcanos. By the mid-twenty-first century, if emissions continue unabated, carbon dioxide concentrations are likely to be at the level of fifty million years ago – that's before ice sheets were able to form over Antarctica and Greenland, because the Earth was too warm. After another two hundred years, they could hit the level of four hundred million years ago[26] – an apocalyptic place to have arrived in just a few centuries.

When you look at forecasts compiled by computer models, the future for our Swiss alpine glaciers is bleak. Irrespective of our success in cutting carbon dioxide emissions over the coming century, the glacier loss will be very high. It is estimated that more than 80 per cent of their ice will be lost by the end of the twenty-first century[27] – so there'll be no more Haut Glacier d'Arolla, or certainly not as I once knew it. The landscape that I experienced and which inspired me as a twenty-year-old student will be irrevocably changed, and the loss will be incalculable.

## 2. *Bears, Bears Everywhere*

*Svalbard*

As my skidoo slowly nosed its way through the fjord's frozen ridges, stacked like waves across the vast sea ice cover, I was hypnotized by the monotonous drone of the engine; it felt like an eternity had passed. One of the many fat fingers of ocean which poke their way into Spitsbergen, the largest Norwegian island in the magical northern archipelago of Svalbard, Van Meijenfjord had been formed by the flooding of deep, glacially moulded valleys at the end of the last glacial period, about 10,000 years ago, as our great ice sheets melted away and sea levels rose.

My destination was the southern shore of Van Keulenfjord, the next inlet to the south, which meant two laborious crossings of the frozen ocean. Although my face was swaddled in hats and scarfs to protect it from the biting cold, my nostrils could not escape the thick, acrid odour of petrol fumes pumping out of the exhaust of the skidoo that crawled ahead of me. My whole torso was braced against the freezing air, and my right thumb almost numb, locked down hard on the throttle in order to maintain the forward motion across the ice. While my eyes were fixed on the skidoo ten metres in front – a black buzzing creature heaving over the uneven ground – my mind was consumed with the fear that I'd lose concentration and miss some obscured feature in the ice which would capsize the skidoo and toss me out like a rag doll.

Skidoo travel was one of many skills I had to master on arrival in Svalbard in spring, all of which I learnt from a bunch of seasoned Norwegian glaciologists, notably a professor from the University of Oslo, Jon Ove Hagen. A legendary figure on Svalbard, Jon Ove had twinkling blue eyes, an infectious smile and the reassuring manner that comes through having seen it all before. I can still recall the silhouette of his slight frame bent over his skidoo as he guided us through valleys and across fjords and glaciers, the ear flaps of his trapper-style hat dancing in the wind like wild bats riding a storm. From negotiating steep valley sides on my skidoo (the trick was to treat it like a small sailboat, balanced by hanging one's body to one side or the other) to dealing with the sudden appearance of bears from the claggy sea mist – Jon Ove taught me everything I needed to know.

Crossing the icy skin of Svalbard's fjords could be treacherous, and was best performed during a short period in spring when the sun had returned after the long winter darkness, but the fjord ice was still firm enough to ensure a relatively safe passage from one land bridge to the other. Even so, there were perils. Occasionally we encountered sections where the sea ice had shifted and opened up small cracks, the lumpy surface of the ice giving way to pools of water, beneath which you were never sure if there lurked more ice or simply deep cold ocean waters. The response of my Norwegian fellow travellers, when faced with these watery sections, was to *just drive faster* – this applied to encounters with most types of holes, from crevasses in glaciers to areas of rotten sea ice on fjords. As you nervously yanked on the throttle to surge forwards, you might sense the back of the skidoo slumping down into the hole as you crossed it, signalled by a muffled thud, the front tracks just gripping enough to propel you forwards rather than down into the abyss. It was terrifying, but kind of exhilarating. And then there was the ever-present issue of bears . . .

Svalbard is one of the strongholds of the great ice bear – *Isbjørn* in Norwegian, *Nanuk* to the Inuit on the Canadian seaboard, or polar bear to many of us around the world. Its Latin name *Ursus maritimus* perhaps articulates most evocatively its dependence on the ocean as a supply of sustenance. While working in Svalbard, this powerful, solitary beast was never far from my consciousness. Whenever I spied a murky yellow shape stumbling in the mist, or a blanched speck on the far horizon, I would ask myself, is it a bear? Is it moving this way? And I'd squint to see whether it had the characteristic cumbrous body and small white head of the ice bear, or was simply a reindeer. (More often than not, it was the latter.)

Svalbard's fjords are vital corridors for ice bears, the sea ice providing a harbour and resting place for their most important prey, the ringed seal. The bear's long neck and speed makes it well adapted to fishing out seals from holes in the ice, or stalking them while swimming.[1] Of the eight species of bear on Earth, the ice bear is the most carnivorous one, and it's obvious what the prospect of an ice-free Arctic towards the end of the twenty-first century[2] means for the animal. Without the ice, the bears lose a vital platform for travel, and access to their main food source.

Only fifty years ago, the ice bear was hunted indiscriminately, including by tourists taking potshots at them during excursions to the Arctic. Dwindling numbers led to a landmark agreement for the bear's protection in the early 1970s by its main host countries (Canada, Denmark, Norway, the USA, the then USSR). It is also now classed as 'vulnerable' by the International Union for the Conservation of Nature. Thus, while you must always carry a gun on Svalbard, the bear's endangered status means that if you use it fatally, the onus is on you to prove that your life was in danger. Philip Pullman cast them as *panserbjørne* or 'armoured bears' in *His Dark Materials*; the first volume of this trilogy of novels, *Northern Lights*, is set in a rather different Svalbard to the one I know, but conjures up a very familiar magic. For the real ice bears of Svalbard, the only armour they possess is the hunting instinct honed through millennia of evolution, now rapidly becoming useless to them as the sea ice retreats.

I've seen many bears on Svalbard – some far, some close, some far too close. I remember every encounter, every bear, the heady blend of fear and wonder, and always the sensation that I was the intruder in this frozen land. There's something about this magnificent bear that draws you in – its freedom to roam confidently across land and ocean, unlimited by borders. Maybe it mirrors our own will to be self-determining; perhaps that's why we feel so emotionally tied up in its fate.

I once stared into an ice bear's small dark eyes, just a few centimetres from mine, through the flimsy Perspex window of a tiny hut which the bear had tried to enter seconds before. The wooden door ought to have been secure, but was on this occasion attached to its hinges only by frayed orange twine – in the kerfuffle of arriving by skidoo late in the day, I'd cut through a piece of string and then again through my forefinger, blade to bone, blood spurting everywhere. I was with two companions – cheery Welshman Martyn Tranter (my former PhD supervisor) and dry-humoured Brummie Rich Hodgkins. Neither of them could stand the sight of blood, so I was left to perform emergency first aid on my wounded finger. Kept from sleep by Martyn's thunderous snore on the opposite bunk, I'd been alerted by a scratching noise; I'd hopped out of my top bunk, to be met by the sight of the bear, its giant front paws pressed against the hut door. He could have entered easily, but luckily he wasn't hungry. Instead, after peering through the hut windows, he commenced some fairly energetic mating rituals with his equally hefty partner just yards away. I bet even David Attenborough hasn't seen this, I thought!

We often make out ice bears to be cute, fluffy creatures, perfect for greetings cards. I have to say that during the brief moment I stared into that bear's eyes, I saw steely voids – simply the wildness of a top-level predator. It was a chilling experience – that of predator meeting prey – and for once, as a human, the tables had been turned. At that moment even the possession of a loaded firearm offered scant comfort, for would I be able to use it against this majestic creature of the north? I refrained from reaching for my camera dangling on the wall yards away, my legs anchored to the ground trembling with raw fear as I contemplated that this was probably the end of my short life. Fortunately, following their tryst, the bears wandered back out into the mist. We reported their nocturnal visit to the Norwegian Polar Institute HQ in Longyearbyen the next day by radio, via a frustratingly squelchy line – 'You would like to report two *birds*?!' asked a disembodied voice in a Norwegian lilt. Our ordeal wasn't quite appreciated for what it was.

This tiny, rust-red wooden hut of bear fame was my home on Svalbard for many months during the 1990s. A quirky, solitary structure sandwiched between old, grey-flecked glacier moraines and the dark waters of the fjord, it was called 'Slettebu', and was perfectly situated to visit the nearby glacier, Finsterwalderbreen. Named in 1862 after the German glaciologist Sebastien Finsterwalder (*breen* means glacier in Norwegian), Finsterwalderbreen lies in Wedel Jarlsberg Land, in the south-west part of Svalbard; just like the Haut Glacier d'Arolla, it is a valley glacier, but it's much, much bigger than its alpine cousin. The glacier itself is broad, sloping gently up to its steep mountainous backwalls some eleven kilometres from the snout. Sitting in the High Arctic, Finsterwalderbreen is notably always cold, the average annual air temperature here lurking at a few degrees below freezing, typically plummeting to as low as minus thirty degrees Celsius over the winter. Compared to my alpine summers beneath clear blue skies and a melting sun, Svalbard felt like being plunged head first into a bucket of ice-cold water.

My arrival on Svalbard for the first time in 1994 reflected a swing in the attention of scientists towards the polar regions, hoping to understand how glaciers here behaved in comparison to warmer, more accessible alpine glaciers. The archipelago, which is about the size of Ireland, has some 2,000 glaciers covering more than half its land mass.[3] Because of the many months of temperatures below zero degrees Celsius, the ice in its smaller glaciers is almost always sub-freezing, and their bottom-most layers, or 'soles', are frozen solid onto the underlying rock. This means that they can only move by the slow deformation of their ice crystals, since they have no water at their beds to help them slide. Essentially, they must crawl.

Finsterwalderbreen, on the other hand, was large enough to stop itself freezing to its bed. Like all polar glaciers, it did have a surface layer of very cold ice, which chilled down in winter, then warmed a little in summer. But if you bored down through this sub-freezing outer layer, you'd eventually hit a 'warm' core of ice at around the freezing point – rather like a jammy-doughnut sliced in half, it had a hard, crusty layer on the outside and a soft, pliable core in the middle. This kind of glacier is technically known as 'polythermal', because it has many temperatures. But when I first visited it in the mid-1990s, scientists knew little about polythermal glaciers, despite most of the glaciers in the High Arctic being of this type.

You might well be wondering – why should it matter if a glacier is polythermal, rather than frozen to its bed? Surely, a glacier's a glacier? Well, the question preoccupying scientists back in the 1990s was this: given that large Arctic glaciers have a surface crust of sub-freezing ice, could any summer meltwater permeate that cold surface layer to reach the (comparatively) warm glacier bed? It was a surprisingly important subject, potentially affecting *everything* to do with glaciers. Meltwater flowing at the bed of a glacier allows it to move much faster by sliding over that watery layer; and how fast the glacier moves in turn affects how much it can erode its rocky underlay, which affects how it shapes its landscape and whether it produces nutrient-rich glacial flour to sustain life in lakes, rivers and oceans. It also fundamentally affects whether microbial life can exist beneath glaciers – because life needs water. A search for this elusive water and its flow paths was the topic of my PhD – which allowed me to spend three years delving deep into glaciers.

Thus, when I was twenty-one, while many of my friends were ensconced in accountants' offices, law firms and management consultancies, I found myself at 78 degrees north, six hundred miles off the pole, traversing a frozen ocean in pursuit of water beneath glaciers. I couldn't believe it really. As an undergraduate, I'd never considered doing a PhD, let alone becoming a glaciologist. However, during the last few days of my Swiss fieldwork, I'd bumped into Martyn Tranter in the Arolla village. Martyn was a bit of a character on the glaciological scene. Brought up in Ebbw Vale in South Wales, and proud to the core, he was well known for his vast repertoire of one-liners, particularly 'every day's a bonus', which he'd exclaim on joyful occasions – he was easily identified by his oversize blue parka jacket which to this day makes an appearance on every field excursion, having weathered some forty years of use. A few hours after we first met in Arolla, he casually asked me whether I fancied doing a PhD in Bristol – he was involved in a large research project funded by the EU to study polythermal glaciers in Svalbard. I'd nervously but eagerly taken the train down to Bristol for an 'interview', which seemed more like a friendly chat, after which we'd adjourned to the pub and I'd miraculously seemed to have won a PhD place. It was a far cry from the gruelling stages of PhD recruitment today, where candidates far outnumber places and have to endure probing selection panels – a change brought about by the fact that nowadays employers think PhDs are quite useful.

In my day few followed the path of a PhD, and I remember some family members being rather suspicious when I announced that I would be embarking on a three-year doctorate on glaciers: 'Why on Earth would you want to be stuck in the freezing Arctic on top of a block of ice? You hate the cold!' (Fair point – I do.) 'And is that really a fitting career path for a woman, among all those men?' Yet to become a glaciologist was almost an instinctive choice to me – I loved the wilderness and freely roaming mountains, I was fascinated by ice, and my ambitions to date had been to become a farmer, an agricultural mechanic or a forest ranger – none of which had gone down very well. My school had refused me work experience in the sixth form, because they didn't regard farming as a respectable career for a girl; while other pupils swanned off to swanky offices in London, I'd simply stayed at home, a misfit with limited prospects. When the University of Bristol offered me a PhD place to study glaciers in the High Arctic, I was elated – maybe there was hope for me after all.

So there I was, perched on one of the most northerly land masses of the planet, in search of water. Working on Svalbard in spring is spectacular – the intense sensory blast of deep blue skies, and sun glinting off surfaces as far as the eye can see, filled me with a pure exhilaration. This was my favourite time of year to visit. After the fjord crossing and some intricate weaving through moraines, Jon Ove Hagen, Anne-Marie Nuttall (a fellow Brit) and I reached the flat snow-clad snout of Finster-walderbreen. I hopped off my skidoo and started padding around the flat snowy expanse in front of the glacier. My feet landed softly upon the snow, as if it were a bed of feathers. I was mesmerized by this flat, twinkling blanket of white silence, but something seemed peculiar. First, I could hear the faint tinkling noise of running water and second, I noticed patches of ground where the snow was darker, wetter. The air temperature was at least minus 20 degrees Celsius, but here was liquid water – how was that even possible? To explain how ludicrous this seemed – imagine putting a tray full of water into your freezer and leaving it for a few days, only to find that it has not turned to ice.

This water, I noticed, seemed to be emerging from beneath a large sheet of very flat solid ice, dusted over by the thin blanket of snow upon which I was standing. This ice cover I later learnt was called naled ice, a Russian term, but described more literally by the German *Aufeis*, meaning 'ice on top'.[4] It was a specific type of solid ice, often found in areas of permafrost where water continued to flow even in winter from deep springs, forming layer upon layer as groundwaters sequentially emerged and froze over the course of a long winter. Naled had been noted in a number of early scientific texts by Norwegian and Polish glaciologists working on Svalbard.[5] The big question, though, was where did all this water come from? There had to be a steady supply coming from somewhere protected against the bitter cold. The only place that seemed remotely obvious to me was the bed of the glacier – but I needed a clever way to test this out, or rather, I needed the waters to tell me where they'd come from.

The thing about water is that it has a memory – a chemical memory. As it passes over rock, it slowly dissolves it, and chemicals from the rock enter the water. Precisely which of these chemicals are present, and how much of them, will reveal the history of the water. For example, you might learn that the water had travelled a long way, or that it had passed through an environment with a lot of mud, or that it had had limited contact with gases in the atmosphere, having originated deep underground. Water can be forgetful, too; over time, dirt can settle and chemicals get so concentrated that they reach saturation and precipitate, forming solid powders which fall out of the water, such as limescale in your kettle. Using this chemical memory is a bit like forensics in a crime-scene investigation. There was water in front of Finsterwalderbreen in the middle of winter – whodunnit?

It's fairly tricky to interrogate water for clues when you're in the icy wastes – you need fancy instruments which require space, power and sterile laboratory conditions. However, a few nifty tools exist which can be used in even the most hostile of environments to provide basic information about the chemical memory of meltwater. A probe that tests how much electricity can pass through a water sample is particularly handy. Pure water isn't much good at conducting electricity – there isn't really anything within it to carry an electrical charge between the probe's positive and negative electrodes – but water is very rarely pure.

As water flows over a rock surface, or through soil or sediment, small amounts of acid in it – often carbonic acid formed from carbon dioxide dissolving into it – attack and break down the rock over time. On medieval buildings built from limestone in cities like Bath, Bristol and Oxford, you may look up and spot gargoyles with rather horrifying, ghostly visages – the acid in the rain has very gradually dissolved their limestone faces away. Thus, from the rock dissolved into it, water comes to contain positively and negatively charged particles called 'ions' – in essence, these are its memory. These ions form from atoms or groups of atoms (molecules) of different elements. Put a current through water and the ions get super-excited, jostling against each other a bit like people in a rave. In doing so, they pass the baton of current from one to the other, bridging the gap between the probe's two electrodes, so that the water body can conduct electricity. The more of these ions there are contained in the water, the more current can pass.

So what did the chemistry of these waters flowing through the naled ice that spring in Svalbard actually tell me? The first thing I learnt, once I'd located a hole in the surface to drop the electrical conductivity probe down, was that the water could carry an electrical current – quite a good one. Maybe it had had contact with rock at the glacier bed? I'd collected a few samples in tiny bottles, filtering out any sediment, to run through some machines on my return to Bristol. These instruments told me that this curious water was rich in ions which had to have come from rock, providing more compelling evidence that it might have emerged from the ground and possibly the glacier's bed. It was hard, though, to work out how the water had got there; I would need to return in the summer, once the armour of naled ice and snow had melted back a bit.

If spring was a euphoric adrenaline rush of skidoo travel, dazzling snowscapes, well-fed polar bears and biting cold, summer on Svalbard was the downer that came after the party. Misty, dank, grey, hovering around 5 degrees Celsius – this was a different type of cold, which seeped through your limbs like ink on blotting paper the moment you stood still, due to the humidity that derives from the West Spitsbergen Current, an extension of the Gulf Stream, which supplies warm water from the Atlantic to Svalbard's western shores. Getting to Finsterwalderbreen was always an ordeal lasting several days. Three flights, from London to Oslo, then to Tromsø at the top of Norway, and finally to Longyearbyen. Founded originally as a mining settlement and named after the American business-man John M. Longyear, who played a key role in exploiting the local coal fields, Svalbard's main town is nestled into the shore-line of Adventfjord, an inland extension of the larger Isfjord, and hosts a sprawled cluster of prefabricated, industrial-looking buildings and a couple of thousand inhabitants. For us, its most important features were the small general store where we bought supplies of everything from tinned meatballs to rifle bullets, along with the bar where we could enjoy the luxury of devouring pizza and ludicrously expensive beer at the end of a field campaign. From Longyearbyen, it was a long day of cross-ing mountains and frozen fjords by skidoo, or an expensive helicopter ride in warmer months, to reach Van Keulenfjord where Finsterwalderbreen hunkers behind a series of towering moraines. Landing in summer, I always felt as though I'd been dropped like a sack out of the air onto an unknown planet, far from human life, and I'd feel a flutter of panic as I watched the helicopter launch into the wide grey sky, not to return for months.

Slettebu, our little hut, became a safe haven from storms and bears. It hosted a wood burner, three bunks and a simple wooden table and bench for eating – adequate but cosy quar-ters for four people, with one always on the floor at night. You never knew whether you would get on with your fellow inmates or not, and inevitably everyone had times when they struggled with the isolation, the separation from loved ones, the cold and the drizzle. During my first summer there one of my room-mates was a tall researcher called Andy Hodson, who sported a thick beard and shock of black hair reminiscent of a were-wolf. When I arrived, Andy had already spent a month at Slettebu (hence the hair), but had been evacuated out of the field by helicopter when a stove in the hut had caught fire and he'd thrown it out of the window with his bare hands. He had sustained severe burns, which needed urgent hospital treat-ment. Despite this, he'd decided to be flown back in for a second round of field action, but was understandably suffering with pain and traumatic memories. One of his greatest hard-ships was only having one functioning hand – I learnt that summer how to roll cigarettes.

I learnt that everyone behaves slightly differently in the field to back home – it draws out parts of you that you can keep well hidden in civilized society. For me, music was an import-ant coping mechanism, and my Sony Walkman proved a faithful companion through long campaigns. When things were tough, I'd wander out alone, rifle on shoulder, to a nearby pebbly beach. I'd gently lay the rifle on the stones, my back to the land, the direction from which I felt a bear was least likely to arrive. I'd fit the headphones snugly around my ears and press 'play', then move frenziedly in time to the music – a wild, contorted dance connecting with some invisible force that seemed to recharge me. Before I left, friends gave me mixtapes of their favourite songs – the Bristol trip-hoppers Massive Attack featured highly. I'd dance madly to 'Unfinished Sym-pathy' on a tape made for me by my PhD buddy Anne-Marie Bremner ('Brems'). The sharp sting of the intro drums surfed by the triangle-like synth always jolted me out of my grey mood,

temporarily dispelling my confusion over how I could feel so
dismal in such a pristine and beautiful wilderness. Then I'd
sling the rifle back over my shoulder and amble back to the
hut, smiling, as if nothing had happened.

It was always worse if you left a partner back home. This
might sound terrible, but it was easier to try to forget them –
the dull ache of missing someone permeated your days if you
tried to cling on to them in your mind and heart. If you detached
and instead threw yourself into your new life with your field
companions, you always felt better, more in the moment. It
took me a while to learn this trick, and I'd often feel guilty
about the disconnection in the early years. The flip side of the
coin, though, was arriving back home and trying to remember
why you'd felt a romantic impulse towards the partner in question in the first place. I'd attempt to explain my experiences in
the field, describing the intense wildness, the communal mirth,
the adventures and misadventures, but few folk really understood – and to be fair, how could they?

Although summers are fleeting on Svalbard, for a few short
months the sun never truly sets and all living things work feverishly to mate, feed and then move on or hibernate before the
long dark winter. I found the constant whirl of this compressed
cycle of life utterly infectious, and threw myself into my project. At night, calmness descended upon the fjord, the snow
and rocky terrain enveloped by ethereal blue tones as the light
slipped away. Then the sun attempted to set, painting the skies
streaky rose gold, only to continue its long path to rise once
more in the east and begin a new day. By August, it finally
started to slink beneath the peaks which rise on the northern
shores of Van Keulenfjord. The distinctive, chiselled shape of
these mountains is one of the most striking features of the
archipelago, and originally gave its main land mass the name

of 'Spitsbergen' (meaning 'sharp, pointed mountains') by the Dutch explorer Willem Barentsz in the sixteenth century. In Van Keulenfjord these mountains lie like petrified waves surging towards brooding leaden skies out to sea in the west – as if once moving, before someone stopped the clock and they were caught as statues. Multi-coloured strata define layer upon layer of rock of different ages and geology, raised up and tilted, dramatically exposed due to the paucity of major vegetation on Svalbard's rocky shores.

I looked forward to my encounter with Finsterwalderbreen in summer, after it had shed its soft winter snow blanket. What would I find? After the usual stumbling trail on foot through the chaotic moraines to reach the glacier front, I arrived close to the spot where in spring I had noticed water burbling beneath the naled ice. Some of the naled was still there, rotting woefully in the summer warmth, collapsed in places and unstable to traverse. Brutally incised through its stratified body was a wild, raging torrent of water, standing waves frothing menacingly around the bends – the proglacial river of Finsterwalderbreen.

What I hadn't anticipated was where all this water was coming from – two places, in fact, rather than one. Some of it was pumping vigorously from a cavernous, ice-walled channel, its dark, foreboding mouth gaping like a giant fish at the ice edge. I had ventured into this channel back in spring, when it was totally dry, and had gazed in wonder at its water- and wind-scalloped ice walls. But much of the water in the main river was not fed from this huge channel, instead emerging from a curious feature close by – a chocolate-coloured fountain spouting a vertical plume of muddy water into the air. The spout was about one metre in height – I called it the 'Finsterwalderbreen Upwelling' – and it seemed to be emerging from underground. Could this be the source of the water I'd encountered in spring, which froze to form the naled ice?

To answer this question, I unpacked the same toolkit I'd used back then to delve into the chemical memory of water. Amazingly, the electrical conductivity of the upwelling water was almost identical to the water I'd encountered in spring not far from this spot. But I needed to understand why, so every day I thrust a plastic sampling bottle forcefully into the violent spew to collect as much sediment-laden meltwater as I could without tumbling in myself. I would quickly filter out the sediment, so that it could no longer dissolve in the water, thus locking the chemistry of the water to that moment in time. A chemical snapshot.

Back in the lab, what this chemistry told me was astonishing. First, there were traces of sodium and chloride (basically salt) in the water. Rocks don't tend to contain much salt, so the most likely source of this was the ocean, repository of vast quantities of the stuff. Yet the ocean was more than a kilometre away. How had all this salt ended up in waters emerging in front of the glacier? The answer to this came when I revisited some samples of snow that I'd collected by skidoo in spring, which also mysteriously contained sodium and chloride.

Snow forms in a cold atmosphere when water droplets sourced from the ocean condense onto dust or ice particles and freeze. In maritime localities like Svalbard, tiny amounts of salt enter the atmosphere as sea spray, and then end up in snow. Actually, by the time snow forms glacier ice, most of the salt has been leached out, so glacier ice is normally fairly pure. The flaw of this as a theory for explaining how salt had ended up in the upwelling was that when I stood in front of the glacier in summer, I could see neither snow nor ocean. This was strange. Only if I clambered on top of the highest moraines in front of the glacier could I just about see the upper part of the glacier eleven kilometres away, where the winter snowfall continued to survive, melting slowly in the frigid air. Wow! This meant that the upwelling had to be fed from salt-infused snowmelt right at the very top of the Finster-walderbreen catchment. Which brought me to my next question: how on earth had this snowmelt travelled eleven kilometres to arrive at the glacier front, popping up via the upwelling?

My investigations into the murky chemical memory of the upwelling led me to the shiny, brassy mineral known as 'fool's gold' – worthless, but easily mistaken for real gold, as had often happened during the California Gold Rush of the 1840s. Its geo-logical name is 'pyrite', a type of iron sulphide mineral, in which two atoms of sulphur are snugly bound up with one atom of iron, giving the chemical formula, $FeS_2$. Pyrite is named after the Greek word for fire (*Pyr*) because it creates sparks when struck against flint, one of the ways in which Neander-thals created fire.[6] It is an extremely reactive mineral – so much so that archaeologists have had a hard time proving that this ephemeral mineral was indeed struck against flint to create fire hundreds of thousands of years ago, because all traces of the pyrite on the recovered flints have since vanished.

Pyrite can be found in tiny amounts in almost all rock types – it's everywhere. This means that it also lurks beneath glaciers, and since the ice is continuously moving over its rocky bed, through slow grinding the reactive pyrites are liberated from their rock matrices. Pyrite reacts with oxygen in water or air to produce an ion called sulphate, along with acid and some dis-solved iron (which later often transforms into a type of rust). If you find the ion sulphate in a river emerging from a glacier, you know that the water must have journeyed across a part of the glacier bed where ice is moving over rock, and pyrite is being continually released from its lithic abode.

In Arolla, if you recall, the boreholes revealed two types of drainage systems at the glacier bed – fast-flowing channels (a bit like rivers above ground, but with ice walls) and swampy areas (lots of connected cavities, through which water flows at a more sedate pace). The slow-flowing water systems cover a much larger area of a glacier's bed (hence being termed the 'distributed drainage system'), meaning that it is usually here that any rock flour eroded from the glacier's rocky bed comes into contact with water for the first time. This, therefore, is where reactive minerals, like sulphides, first start to dissolve in water, releasing sulphate ions. By the time the rock flour makes it into the fast-flowing channels, its sulphides have often been exhausted, and there is little new sulphate added to the meltwaters. So here's a marker for the two different types of plumbing systems beneath glaciers – sulphate ions.

When I analysed the water that I'd collected from the bubbling upwelling at Finsterwalderbreen, I was amazed to find that it contained a lot of sulphate ions. When I looked at the waters that I'd bottled in spring, seeping up from beneath the blanket of naled ice, I found the same thing – a lot of sulphate, plus ions that could only be generated by water dissolving rock. So those mysterious upwelling waters were in fact snowmelt which had journeyed slowly through an environment where a lot of rock was being ground up and most likely via slow-flowing drainage systems, giving lots of time for rock to react with water.

But why did meltwaters emerge via an upwelling in the first place? This is a question for which no one yet has the definitive answer, but it would appear that it's hard for meltwaters flowing at the bottom of a glacier like Finsterwalderbreen to pass its cold snout, which is well and truly frozen onto the underlying rock, and acts a bit like a dam. This difficulty of waters escaping the bed at polythermal glaciers is common. At John Evans Glacier on Ellesmere Island in the Canadian High Arctic, because meltwaters can't escape the frozen snout, they build up in the glacier in early summer until they reach such a pressure that they literally blast a connection through the snout, also exploding vertically through the glacier surface to create an artesian fountain, a bit like a whale blowing for air.[7]

By accident, I'd encountered a similar phenomenon at Finsterwalderbreen. This was 1995, towards the end of a long field season, and I was weary, longing for home comforts and food that wasn't crackers and tins of texture-less meat or fish balls. With about two weeks to go, my companions and I had headed to the glacier to carry out our daily check of the instruments installed in the main river to measure its depth (and therefore discharge), electrical conductivity and sediment levels. Suddenly we heard a muffled roar, then a rumbling crash, and minutes later the water level in the river was rising fast and icebergs the size of people were being tossed down the channel. To my alarm, I glimpsed a mass of cables being yanked over the steep riverbanks as the mechano-like structure of angle iron and poles fixing the instruments in place was dragged mercilessly into the rising torrent.

As I sprinted in panic to save our station, I tripped over a boulder protruding from the mud and fell down hard, my right knee smashing into the rock over which I'd stumbled. Agonizing though it was, there wasn't much option but to just carry on. (It later emerged that I'd fractured my kneecap.) Now lacking any instruments to measure the river discharge during this rare event, we rigged up a simple vertical pole, with five-centimetre gradations of height along its length, in a slack water area of the river. For the remaining two weeks of the season I camped in a small tent next to this lone pole, crawling out of my canvas abode and into the dirt like a wounded animal every few hours on my hands and one knee, to read off whether the river level was rising or falling.

What these hard-won measurements told me was that the floodwaters, which continued to emerge from the glacier for several days and had cost me a functioning kneecap, probably came from a giant lake beneath the ice.[8] Over the summer this lake had filled slowly, then catastrophically failed, releasing its water as a torrent into the channel that fed the giant ice cave – the channel that I'd thought didn't connect to the glacier bed at all. So there were two ways water at the bed of Finsterwalderbreen bypassed its frozen snout – as a steady flow of melt via a secret underground passage that ended with the bubbling upwelling, and as a boom-bust river that tracked the western edge of the glacier and terminated in the ice cave. Based on painful experience, I preferred the upwelling.

My fellow researcher Anne-Marie Nuttall, of the Scott Polar Research Institute in Cambridge, also taught me a great number of practical skills in those early years. Her measurements of the ice flow at Finsterwalderbreen subsequently confirmed my theories about the water at the bed of this glacier – her careful surveying of the position of aluminium stakes drilled into its surface showed that the ice sped up from about ten metres a year in winter to thirty metres a year in the glacier mid-section in summer.[9] Despite its cold surface layer, meltwater was clearly able to reach the bed of Finsterwalderbreen, greasing the glacier sole and enabling the ice to slide more quickly downslope.

It may sound so simple – summers spent ambling around a glacier and its lunar-like forefields, filling tiny clear plastic bottles with filtered meltwater, putting them in a box and bringing them home to run through various machines in the lab. But honestly, I spent the first year of my PhD not really having much of a clue what I was doing or why. One week I was

togged up in a gown, mortar board perched on my head, enduring a rather pompous graduation ceremony in Cambridge, and
the next week I was on a plane bound for Svalbard, lugging a
brand-new rucksack that I could barely lift, complete with
gleaming ice axe, crampons and newly acquired layers of fleece
and Gore-Tex. (If I'd learnt one thing since Arolla, it was that
glaciers were *cold*.)

Of all the places I might have been dispatched to, there was
something about Svalbard which captured my imagination
even before I'd boarded the plane. All those family holidays to
the Cairngorms when I was a child had awakened in me a fascination with what lay 'north'. Every year, as we sped up the
motorway from London, I'd notice the big blue sign with white
writing that announced, 'THE NORTH'. It always sent a
shiver of anticipation through me. So when I finally set off
for Svalbard – the real north – I could barely contain my
excitement.

I quickly realized one key thing about fieldwork – if you
think you are there to work, you're gravely mistaken. You're
actually there to survive, and perform some research along the
way – if you're lucky. The business of survival included dealing
with changeable weather (not least in one's mood), finding
ways to make the bland rations taste of anything other than
reconstituted cardboard (tins of fish balls, or 'fiskeboller', were
the main focus of creative cookery – when not needed for rifle
practice on the beach), collecting driftwood for the stove, and
washing clothes and oneself in icy cold water. Once a month,
I'd try to wash my slowly matting hair – plunging it into an ice
bath from the fjord, which made my skull feel like it was shrinking to diminutive proportions. Days were long, often punctuated
by some disaster either in the weather or in some piece of kit
that had failed to work or got washed away for the umpteenth

49

time – Rich Hodgkins always put things into perspective at the end of an exasperating day. 'Well – no one died, Jems,' he would remark in his soft, deadpan tones. In some ways I found all this 'surviving' a grounding process, and both body and mind relaxed as I returned to a relatively simple existence that was perhaps closer to that of the earliest humans.

However, polar bears were a constant source of anxiety. Their sighting on land has risen over the past few decades as the sea ice around Svalbard has diminished, and they've been forced to find food from onshore – bird eggs, geese, harbour seals, walrus.[10] By midsummer, the sea ice had retreated to the far north-east of the archipelago, away from the temperate Gulf Stream to the west. Bears tend to follow the seals, and therefore the ice – so if you see a bear in western Svalbard in the summer, you know that it is probably hungry and seeking a snack on its way east. Your next thoughts are, in rapid succession, has it seen me? Should I load the rifle? How far am I from safety? We deployed trip wires around the hut as an early warning system against bears – something running through them triggered a series of loud explosions. One problem, though, was that the wires were pretty much invisible. I lost count of the times I walked right through them, only to be scared rigid by a deafening bang above my head, shortly followed by the arrival of my companions, rifles at the ready. 'Sorry, sorry, sorry,' I would cry apologetically, as their faces registered a mixture of relief and irritation.

Early one morning I was happily making porridge while everyone else was slumbering either inside the hut or in over-flow tents outside. Suddenly, a huge bang pierced the silent morning air and reverberated around the steep fjord walls, shortly followed by a loud crack. The first was caused by a trip wire, the second a rifle shot from one of our two German

Mauser .308 rifles (the year 1945 eerily engraved on its shaft). An adult female bear had wandered into camp with a cub in tow. Dave Garbett, our field assistant, hearing some agitated scuffling outside his tent, had sleepily half-unzipped his tent door to see the mother lurching towards him. He'd grabbed his rifle, ripping the tent canvas in the commotion, to fire over the bears' heads in warning. I happened upon this scene, wooden spoon gripped tight in my trembling hand, and watched speechless as the startled pair of bears charged clumsily through camp, before finally heading for the fjord.

When I was not worrying about bears, my main concern during my later visits to Finsterwalderbreen was this: given the existence of water at the glacier bed, could it support life? Sunlight sustains most life on Earth today – through the process of photosynthesis, plants use solar energy to combine carbon dioxide with water and make organic molecules – initially glucose, but ultimately also proteins and fats. As humans, the nutrients we consume ultimately derive from plants – we eat them, we consume the cows and other animals that eat them, and so on – it all comes back to the plants and therefore the sun.

But beneath glaciers there is no light, only crushed rock and (as we now know) water – so how might life survive in this deep, dark underworld? This question took me back to the iron and sulphur bound up within pyrite (fool's gold) – both these elements have been present on Earth since almost the very beginning of time, several billion years ago. Scientists believe that a certain type of microbe called a 'chemotroph', which could use chemical energy instead of light, flourished on ancient Earth. That's because many chemical reactions on Earth release energy, and some microbes have found a way to tap into it; for example, when pyrite reacts with oxygen, perhaps at the base of a glacier, the sulphur within the rock is converted to sulphate by means of an 'oxidation reaction', which generates energy that a microbe can take advantage of, to survive and grow. Essentially, the microbe consumes the rock.

One thing I had a burning desire to know, which might help tell me how the microbes were surviving, was how heavy the sulphate ions in the upwelling that spouted from the forefields of Finsterwalderbreen were – which may sound like an obscure line of enquiry, but atoms of some elements, like sulphur and oxygen (which make up the ion 'sulphate', $SO_4^{2-}$), can have different masses. On Earth there are two main types of oxygen atom: one is heavy, one is light. These are known as 'isotopes', from the Greek words *isos* (equal) and *topos* (place); they have different masses but are the same element, so they occupy an identical position in the periodic table.

The oxygen in the water molecules of glacier ice tends to be light. This reflects the journey that water molecules take before they settle on a glacier as snowfall. If you imagine that water in the ocean contains a lot of light and some heavy molecules of water, when this water evaporates, it is easier for the lighter molecules to leave the ocean and enter the air. As the moisture-laden air travels to the peaks and the poles – the heavy oxygen atoms are also lost more readily in rain or snow closer to their source in the ocean – it is much easier for the light stuff to make it high onto the glaciers. Oxygen present as a gas in our atmosphere, on the other hand, is very heavy – mostly because when plants and animals use oxygen in air to oxidize organic carbon to generate energy, they preferentially take up the light oxygen in the air, leaving heavy oxygen behind. Plants on the land and in the oceans help compensate a little for this, as they take up water (which is relatively light) and return some of its oxygen molecules to the atmosphere. In Arolla, scientists had already reported that the oxygen in the sulphate in glacier runoff was light, which meant that the sulphur in the pyrite could not have been converted to sulphate using atmospheric oxygen; instead it had to come from the light glacial meltwater.[11] The only way this could have happened was with the help of a clever group of microbes, which are able to use a form of iron (a bit like rust) to oxidize the sulphides. In doing so, they use chemical energy released during the reaction to take up carbon dioxide from the meltwaters and produce organic molecules for their cells. Bacteria such as these may have been present on early Earth – and today, in the Alps, they can be found living at the bottom of glaciers.

I ran some tests on water from the Finsterwalderbreen upwelling to see if things were similar there. On the contrary, the sulphate here had both heavy oxygen *and* heavy sulphur (which also has light and heavy forms).[12] This was strange, and could only happen if another type of microbe, a sulphate-reducing bacterium, was taking the sulphate and using it as a way of oxidizing organic carbon to get energy. These particular microbes are what we call heterotrophs – from the Greek *hetero* (other) and *troph* (nourishing) – because they must rely on food produced by others. They thrive in places where there's no oxygen; you'll often find them in landfill sites. The sulphate-reducing bacteria prefer to consume sulphate containing light sulphur (which they convert to hydrogen sulphide gas – the rotten-eggs smell you also find in landfill) and oxygen, leaving some or all of the sulphate with the heavy sulphur and oxygen untouched. These discoveries revealed that the meltwaters upwelling at Finsterwalderbreen came from an environment which was much more starved of oxygen than in Arolla; they also showed me that some microbes prospered under such conditions.

On Svalbard, mostly by filling tiny bottles with meltwater,

I discovered many things. I found water flowing at the bed of an Arctic glacier taking unexpected and explosive paths to bypass the frozen glacier snout. And I found evidence for active life in this water – microscopic organisms which were adapted to survive in the cold, without oxygen, living off energy produced by chemical reactions. This was a revelation; it suggested that the meltwater produced over millions of square kilometres of glaciers across northern Canada, Scandinavia, Greenland and the Russian Arctic was not just running straight off the ice, but plummeting to the depths, where it greased the slide of the polar ice over its bed. This explained how the glaciers were moving, grinding their rocky platforms and feeding giant ice rivers with fine fertile sediments. It also meant that a large area of the planet beneath glaciers, previously assumed to be essentially dead, was very much alive. We could no longer think of glaciers as frozen, sterile wastelands – they were as much a part of Earth's biosphere as the forests and the oceans.

My journeys to Svalbard ended in the year 2000, but this ragged-edged archipelago will for me always be a magical land, with its eclectic mix of glaciers ranging from those that end abruptly on land to those which mysteriously creep into the ocean, and its rich, theatrical cast of wildlife – the ice bears free to roam, the beluga whales that swim the fjords sloughing their outer layer of skin in the shoaling waters, the Arctic foxes furtively sneaking through camp in search of sustenance, and the Arctic terns dive-bombing those who dared cross their plains. This place is for ever lodged in my heart, like a boulder in a riverbed. I often wonder, doubtfully, whether it would live up to my vivid memories if I went back now.

I've not seen the changes that have affected Svalbard's shores due to our warming climate, but I do know they are happening. The Arctic has warmed at more than double the rate of the rest of the world over the past two decades.[13] The reason for this polar superheating is complex, but involves what is known as (somewhat ironically) a 'positive feedback effect' – in other words, the warming brings about a change that then causes more warming. For example, Arctic warming has been closely tailed by the retreat of sea ice across its central ocean basin, the Arctic Ocean. Sea ice acts as a kind of air conditioner for the Arctic, since its white surface reflects much of the sun's rays back to space. Since warmer air can hold more water vapour, which is readily supplied from the now extensive open ocean expanses, the greenhouse effect of the additional water vapour also amplifies Arctic warming.

These days the skidoo journey across the frozen fjords would undoubtedly be more challenging than it was in the 1990s. Is-fjord, the main fjord close to Longyearbyen, was named for its high ice cover, but has not fully frozen for a decade. This is the effect of warm, salty water from the Atlantic, ferried by the West Spitsbergen Current, now intruding closer to the surface along Svalbard's coast.[14] This warm water imports nutrients and food for plants and creatures that live in the surface ocean and which then provide sustenance for larger forms of sealife like fish and shellfish. This may help to support greater populations of mackerel, cod, haddock and capelin to the north of Svalbard in coming decades.[15] But it has caused winter sea ice to shrink by 10 per cent per decade since the late 1970s,[16] and fjord ice with it,[17] both prime hunting grounds for the great ice bear.

The invasion of the warm water has also been felt keenly by Svalbard's many glaciers whose tongues float on the ocean and melt faster in warm ocean water – even landlocked Finsterwalderbreen has retreated a whole kilometre since the 1990s – reflecting hotter air temperatures. Some of the smaller glaciers on Svalbard are now so thin that they are no longer polythermal and are frozen to their beds, unable to slide; warming has, ironically, caused them to cool. I often ask myself, would I go back to Svalbard to study these drastic changes, even if it risks contaminating my precious memories? The answer is yes, and I am sure that one day it will happen.

For now, I'm happy to be reminded of wild times on Svalbard in a completely different way. Some years after my adventures there, when I reached forty, I decided that I needed to learn something new, and traded my climbing boots for a four-legged challenge. I'd never spent time with horses, always preferring to roam the remote fringes of the planet without company. It took me a few months to learn to ride on the flat before I graduated on to rounds of show jumps and heady gallops around the countryside. Then I fell in love with a beautiful dark horse, a mare called Usher. She was smart, talented and wouldn't stand for any nonsense – all things I respected, particularly the latter, which I would have loved to be better at myself. Often she'd catapult me to the floor when she didn't feel in the mood for control and dressage manoeuvres – I kind of understood her distaste. At other times she'd charge round a course of three-foot jumps with crazy abandon – breaking my bones on more than one occasion. No one else wanted to ride her, so she became mine.

One sad day she fell lame and we could have adventures no more. She underwent surgery, partially recovered, but our former relationship based around a kindred pursuit of freedom did not. We've since begun a different journey. In April 2018 she gave birth to a son – a stunning bay colt with a white star splashed across his forehead, like a retreating snowpack against its dark, rocky surrounds. His name is Finsterwalderbreen, or 'Fin' for short.

# PART TWO

# The Great Ice Sheets

# 3.   *Plumbing the Depths*

*Greenland*

I snatched the headphones dangling above me just in time to buffer my eardrums against a rising cadence of metallic roaring as the blades above my head whirred faster and faster until they were a blur. The helicopter wobbled its hefty torso, clumsily raising its feet one by one off the ground until it hovered upon a pillow of swirling sand, its downdraft vigorously scouring the moon-like proglacial plain of Leverett Glacier, a large land-terminating glacier in south-west Greenland. As the pilot pulled hard on the joystick to surge forwards, banking flamboyantly on a wide arc to tilt alarmingly as we powered our way skywards, I smiled to myself as I mentally dislocated from the Earth. It was a relief now to be past the parade of faff that always preceded such excursions, as gear was stowed carefully in the hold, flimsy items on the ground which might become airborne were secured to the ground, and meanwhile the pilot paced around, stroking his beard, tut-tutting that we were over the weight limit. These moments of leaving, rising to the skies, were a feature of working on Greenland, and I revelled in the feeling of freedom.

Buzzing around in helicopters, negotiating pot-holed dirt tracks in 4 x 4 off-road vehicles, crossing raging rivers in rubber dinghies – everything about these expeditions was huge. Indeed, that's the first thing you notice about Greenland – its immensity. The land mass is about the size of Mexico, but

covered by a thick dome of moving glacier ice several kilometres deep in its centre. When I first arrived in Greenland, its
soft, ice-moulded hills seemed familiar to me – similar to those
of the Cairngorms. Yet in the other direction lay the ice sheet –
an outrageously flat, gleaming layer of white as far as the eye
could see. I realized that this was what Scotland must have
looked like 20,000 years ago, when the British Ice Sheet blanketed much of our mainland. In the local Inuit tongue, this
vast area is known as *sermersuaq*, 'the great ice'.[1]

The story of how this ice-encrusted island was named
'green', rather than white, dates back to the Norsemen who
colonized parts of west and southern Greenland. Erik Thor-
valdsson (often known as 'Erik the Red'), one of the early
Norse settlers, arrived from Iceland with a fleet of longships
around AD 985 at an inlet on the southern tip of Greenland
called Eiriksfjord (now named Tunulliarfik Fjord). His people
are commonly known as Vikings, from the Old Norse *víkingr*,
meaning 'to go raiding' – though not all Norse were raiders,
and this was certainly not how they referred to themselves.[2]
Erik's father was a chieftain in Norway, but both men are
thought to have fled due to inter-clan feuding, which also
caused the dispersal of many Norse to Iceland, the Faroes,
Shetland, Orkney and the Hebrides during these times.[3] Erik
called his new home 'Green Land' because along its margins
lay a seductively lush, vegetated land suitable for farming, and
he wanted to create a welcoming image of this far-off land so
that settlers might follow him.[4]

From the start, I found the scale of Greenland's ice sheet
quite daunting – there were so many enormous questions
which almost seemed unanswerable – but it was this challenge
that drew me in. My arrival there in 2008 followed a series of
'dry' years in research funding, when I'd repeatedly failed to

win grants to continue my work in Svalbard. Valley glaciers were no longer considered quite as cutting-edge to the research council funders; instead, glaciologists had become obsessed with the vast ice sheets that contain just over two-thirds of the Earth's freshwater.

Greenland is of particular importance to the British Isles, connected as we are to this gargantuan lump of ice via the winds that bring Arctic blasts in winter, as well as the deep, cold currents originating in the Greenlandic seas that nose their way south down the Atlantic past the Faroes and Shetland. These currents are an essential part of a conveyor-belt system of ocean currents called the Atlantic Meridional Overturning Circulation (known more catchily as the 'AMOC'). This acts like a heat exchanger, moving warm waters to the chilly north and cold waters to the balmy south. Its northward currents, the Gulf Stream and North Atlantic Current, are one reason why the British Isles experience such mild weather – if it were not there, the air temperatures would be up to nine degrees Celsius cooler.[5] Thus the future of the UK is tied up with the fate of Greenland and its ice sheet.

Technically speaking, to be considered an 'ice sheet', and not just merely an ice cap or glacier, a body of ice has to be larger than 50,000 square kilometres, and blanket the mountains. Our planet has just two of them at present – Greenland in the northern hemisphere, and the adjoining East and West Antarctic Ice Sheets (considered together as the 'Antarctic Ice Sheet') in the southern hemisphere. Antarctica's ice sheet, at almost fourteen million square kilometres, covers an area about seven times that of the Greenland Ice Sheet. In much colder times, such as the peak of the last glacial period 20,000 years ago, there were several other ice sheets, including the Laurentide Ice Sheet covering North America and the Eurasian ice sheets in Europe, of which the largest was the Fennoscandian Ice Sheet centred over Scandinavia. Even Britain had its own ice sheet, appearing about two and a half million years ago as the climate cooled following the Pliocene warm period[6] – it later became connected to the Fennoscandian Ice Sheet next door. During these chilly times, the climate was a bit like that of the low Arctic today, in places like southern Greenland, and woolly mammoth and woolly rhinoceros roamed the tundra beyond the ice in southern England.

The climate warmed rapidly at the end of the last glacial period after the Last Glacial Maximum around 20,000 years ago, following small shifts in cycles of the Earth around the sun, which raised the heat supply to Earth's surface. These small changes in heat were amplified by feedback processes, as greenhouse gas-producing bogs and wetlands expanded in the wake of retreating ice sheets, and shrinking ice caused a drop in the Earth's surface reflectivity. Thus, the Laurentide and Eurasian Ice Sheets wasted away, and the Greenland and Antarctic Ice Sheets retreated. The loss of these great ice sheets, which covered nearly one-third of the Earth's land surface at the time (compared to about a tenth today), caused sea levels to rise by up to 120 metres as their meltwater ran into the oceans over the course of about 10,000 years, with most of the increase happening between about 16,500 and 8,200 years ago.[7] This rise in sea levels equated to about one metre every century on average – not far off one of the worst-case scenarios for the predicted sea-level rise by the end of the twenty-first century, which is a little less than one metre.[8]

Confusingly, though, the sea-level change was different everywhere; the northern part of Britain's land surface where the ice had been thickest, for example, started to rise as its ice wasted away and the removal of all that weight caused it to bounce back. (Rather like pressing into a piece of foam to create an indentation, but once you remove your finger the foam pings back out again – this is what happened to Scotland.) The opposite occurred further south; here, the land surface fell as its underlying mantle shifted back to fill the hollows where ice sheets had once been, and as more and more meltwater sloshed into the Atlantic and added extra pressure, squashing it and causing the land to fall relative to the sea. So, Britain was a bit like a see-saw – Scotland going up, southern England going down. In fact this is still happening 20,000 years on, by a few millimetres every year. It means that in the south, any further rise in sea level due to glacier melting will be even more keenly felt.[9]

The Greenland Ice Sheet itself is a giant dome of ice which grew to a large size by two to three million years ago. It hosts hundreds of glaciers that protrude as white fingers around its perimeter, transporting ice downhill through deep glacier-carved troughs towards the oceans. These icy digits are called 'outlet glaciers' – some of them terminate abruptly on land and supply melt to wild, turbulent rivers, and others have tongues which float in the ocean, 'calving' bergs into fjords. At the start of the 2000s there was only a hazy understanding of how the Greenland Ice Sheet worked – for it was so huge that the kind of experiments which were perfectly feasible to carry out on a valley glacier just weren't practical. You couldn't walk from the edge of the Greenland Ice Sheet to its middle – it would take you about a month and you'd need to traverse some fairly deadly terrain. So no one really knew how it was plumbed, how this affected the flow of the ice, or how fast the ice sheet was melting. Like Finsterwalderbreen in Svalbard, most Greenlandic outlet glaciers were of the polythermal type – with an outer layer of cold ice, and a core of warmer, more pliable ice. The only difference here was that the ice was up to several kilometres thick and the surface area of an entire nation – so if something happened to it, the effects would rico-chet around the world.

The Arctic has been warming at more than double the rate of the rest of our planet over the past two decades – and its oceans are following suit.[10] This does not bode well for Green-land's ice sheet, which locks up about seven metres of possible sea-level rise. The 'health' of big ice sheets like Greenland's relies on sufficient snowfall to balance against surface melting, plus the loss of icebergs (by 'calving') and melting of ice tongues which end in the sea – surface melting and iceberg calving each account for about half the ice sheet's mass loss every year. The problem for Greenland is that while snowfall is not changing very much, the surface and tongues of the ice sheet are fast succumbing to the warming of both the air and oceans around them. Thus, the amount of summer melt on Greenland has been rising since the 1980s as its climate warms, with the three highest melt years on record observed in the last decade, in 2010, 2012 and 2019.[11]

The retreat of Greenland's glaciers that end in the sea (called 'tidewater glaciers') due to warming of the oceans has been an alarming trend of recent years.[12] Shrinking up their fjord chan-nels as ocean waters thaw their frozen tongues, tidewater glaciers react by moving ice more quickly from inland to replace icebergs and meltwater lost off their lower trunks, in turn depleting their frozen reserves.[13] As the tidewater glaciers get thinner, their surfaces become lower and melt more easily in warmer air – the vicious circle also known as a positive feed-back effect. Staggeringly, Greenland's tidewater glaciers have receded by over one hundred metres a year on average between the years 2000 and 2010.[14]

Already the effects are starting to be felt far away. Greenland's melting ice sheet is now understood to be the greatest glacier contributor to global sea-level rise, outpacing both the Antarctic Ice Sheet and the numerous small mountain glaciers which had previously dominated glacier contributions to the rise in sea levels.[15] This sea-level rise created by a shrinking Greenland Ice Sheet is just over two-thirds of a millimetre per year (0.77 mm) at the moment,[16] but it is accelerating and we don't know quite where it will end.

On 1 August 2019, Greenland lost more ice in one day than ever recorded before – about twelve and a half billion tonnes of water flowed into the sea, (very) roughly equivalent to a giant swimming pool about the size of greater London, some eight metres deep.[17] As Greenland melts, its neighbouring oceans are becoming fresher, and it's feared that all this freshwater could slow the AMOC heat exchanger which evens out temperature differences that naturally occur between the poles and more temperate latitudes. It does this by instigating a conveyor belt of ocean currents which span the Atlantic Ocean. Ocean surface water in the Nordic and Labrador seas on either side of Greenland cools until it freezes to form sea ice. Sea ice does not easily incorporate salt into its crystal structure; instead the mineral is rejected into the nearby ocean water, making it dense and salty. These cold saline ocean waters sink and edge south along the ocean bottom. Countering this deep southwards flow are currents of warm surface waters moving north via the Gulf Stream which extends into the North Atlantic Current, driven by powerful winds. Together, the cold currents moving south and the warm currents moving north form the conveyor belt – if you change a part of that belt, then the whole AMOC shifts. Reduced sea-ice formation and a greater quantity of freshwater coming from a melting Greenland are believed to be disrupting the sinking of dense, salty water in the Arctic seas, weakening the AMOC, which in turn might lead to stormier, colder weather in Europe.[18] Scientists think that the AMOC is unlikely to collapse entirely, as it has done during big melt periods of the past, when entire ice sheets melted – but there is much uncertainty.

When I arrived with a small team in Greenland in 2008, we had a single mission. We wanted to find out whether the vast volume of meltwater produced on the ice sheet's surface every summer flowed into and within the ice sheet, and whether it greased its sole – for if this were the case, more warming might cause the ice to speed up and run faster into the oceans, affecting sea levels, currents and marine life. I had set myself a monumental challenge – the glacier which had caught my eye in Greenland, Leverett Glacier, was more than ten times larger than Finsterwalderbreen in Svalbard, albeit a tiny fragment of the giant ice sheet. The quest commenced with fifteen kilograms of bright pink rhodamine dye, poured into a large moulin far beyond the ice margin at Leverett Glacier – this was the purpose of that first helicopter ride. I'd had numerous long conversations with my old friend Pete Nienow from the Arolla days about how to attempt this feat. The general principle was that if a clear channel connected the moulin to the front of the ice sheet, the dye would eventually show up in the main pro- glacial river – the handy thing about this dye being that using a fluorometer you could detect it in tiny quantities in the water, even when the water was no longer pink.

So there I was, huddled in the back of a tiny helicopter, star- ing down at the lumpy ice surface, riddled with snake-like turquoise meltwater streams and flat, glassy lakes and ponds. Using a combination of GPS and my eyes, I was desperately trying to spot a giant moulin which had been reported about fifteen kilometres inland from the edge of the ice sheet – not as easy as it might sound. One of the disconcerting things about flying across an ice sheet is that everything looks pretty much the same – a collage of white, turquoise, black and grey, seemingly stretching to infinity. The air that day was cool, probably only a few degrees above freezing, but sufficiently warm to permit the disintegration of tiny ice crystals at the surface, blasted by the sun's powerful rays, creating a life-sustaining supply of melt. The midnight sun in Greenland is unrelenting, with twenty-four hours of daylight and the high reflection off the ice. Our brand-new tents started the field campaign a garish orange, in glaring contrast to the drab grey sandscape of the proglacial zone; within three months they had faded to a pallor that rendered them near-invisible (unlike my face, which soon acquired a deep tan).

Yet the ice wasn't as pristine as one might have imagined. A combination of abundant light and water make the surface of the Greenland Ice Sheet an inviting habitat for colonization by tiny microscopic life forms like algae. They create mats which cling to the surface and grow larger as the melt season progresses, forming darkened areas on the white ice as they spread.[19] These algae have pigments in their cells, from brown to deep purple, which act as a kind of screen protecting them from the sun's harmful ultraviolet rays. This may be useful for them, but it's bad news for the ice sheet, because the darker algae covering the ice absorb more of the sun's rays than bright white ice, so it melts faster.[20] This is another of those 'positive feedbacks' – more meltwater on the ice surface, which encourages algal life, thus causing surface darkening, absorbing more of the sun's rays, which triggers more melting.

It took a while to track down this elusive moulin. Eventually from the helicopter I spied an enormous aquamarine channel, meandering like a serpent, its clear waters illuminating dark lines of crevasses and fissures in its cold icy base – a mesmerizing sight. Where had it come from? And where was it going? We followed the river downstream for a while, the helicopter tracking its meandering form like a fugitive on the run, James Bond-style, until suddenly the river just vanished. As we hovered above, I peered out of the steamy window to gasp at the most gigantic hole I'd ever seen, a cavernous void which effortlessly gulped down the crystalline contents of the river into its cool blue vortex, like a giant plughole in the bottom of a sink. That'll do, I thought.

But how to empty three bulky containers of dye into a vast ice-walled hole some tens of metres deep? In Arolla, the task was much easier – you only needed a few hundred grams of dye and the moulins were tiny in comparison. We landed the helicopter, and I sauntered over to the edge of the precipice, feeling rather dizzy as I gazed into the gaping hole. Hmmm, that doesn't look so simple, I told myself. I suspect the most sensible response to this situation might have been to try to find a more accessible moulin, but no, I'd come all this way armed with fifteen kilos of dye, and I was determined to make this one work. So I sunk some ice screws into a patch of solid-looking ice and shouted to my research assistant, a strong and similarly stubborn character from the Czech Republic: 'Marek, if I tie into this rope, can you belay me down the hole so I can dump the dye direct into the water as it goes down?' And that is what we did.

Less than two hours later, our team at the ice front called on the satellite phone to say that the dye had emerged as a single sharp burst – which indicated that large rivers of water must be flowing fast beneath a kilometre-thick ice sheet, delivering copious volumes of icemelt to the glacier front. Over the following Greenland seasons we repeated this endeavour many times at other moulins, gradually moving further inland from the ice edge, using some more sensitive gas tracers to see what happened to moulins up to sixty kilometres from the ice margin. Nobody had managed to trace water through an ice sheet before, and it felt incredible to be pioneers.

This work may have delivered huge bursts of adrenaline and euphoria, but it also entailed challenging stays in remote field camps for months on end. Establishing a remote field camp at the edge of the Greenland Ice Sheet was not without its difficulties. There were months spent packing several tonnes of science equipment, then there was the mad dash in a van to Denmark to secure its place on a cargo ship to Greenland, followed by the long wait while the ship sailed to the main town of Kangerlussuaq, the frustration when all the equipment arrived a week or two late, then the arduous task of repacking it to fit into a sling (basically a large net) dangling beneath a helicopter. Helicopters are expensive to hire, costing about £2,000 an hour for the smallest one – and for this reason, we only used them for putting in or taking down a camp, or for long forays into the ice-sheet interior to hunt for moulins. If we were lucky there was a helicopter stationed in Kangerlussuaq, just twenty kilometres from Leverett Glacier. If we were unlucky it had to come from Nuuk, about an hour and a half away, which bumped the bill up by about £6,000, unless there was a storm and the helicopter had to turn around and try again the next day, by which time we would be into serious budget panic territory.

When stationed at our tiny camp – a jaunty collection of sleeping tents, a couple of larger laboratory tents and a big mess tent for cooking, hanging out and the occasional party – we had to opt for cheaper means of bringing people in and out. This had its hair-raising moments. The first part of the journey involved a twenty-kilometre cycle ride on a bumpy dirt road or across shifting sands, usually with a fifteen-kilo rucksack. It was slow going, and by the end of it one's legs were in agony. Next, after we had hidden the bikes behind a boulder, there was the 'river crossing'. In order to get to the edge of Leverett Glacier, we had to cross a turbulent, foaming river that flowed from an adjacent glacier. To achieve this goal, we rigged up a rubber dinghy on a pulley system across the fifty-metre-wide channel in one of its 'calmer' sections. You would climb into the dinghy, steady yourself, and then wait for four people on the opposite bank to drag you across on a rope, the boat tossing and turning in the standing waves at the channel centre as you clung white-knuckled to its bulwarks. Then it was a ninety-minute slog over the sinking sands and stony channels of Leverett Glacier's expansive flood plain, before a leg-burning climb up and over a rocky hill to reach the final destination, known as 'Camp Famine' (a term coined by Martyn Tranter, who was now running a camp on the glacier next door, 'Camp Paradise', which was much closer to civilization), due to its apparent desolation and remoteness – which meant that we often ran out of tasty food by the late melt season.

People often ask me what it's like leading teams of scientists to remote places – how do you manage *as a woman*? My answer to which is always that, to me, it feels the most natural thing in the world. In the wilds it's a level playing field; it doesn't matter if you're a student or a professor, you all have to pitch in. For me, the most natural way to lead has always been from the ground. Sometimes this means people see your vulnerabilities, as you struggle to lug a pack of thirty kilos across the tundra, or stagger out of your tent barely able to utter 'Good morning' after not sleeping a wink. As a leader you need to have a goal, of course, as well as a plan for achieving it and a way of getting people behind you – but sharing ownership of the goal has always felt to me the obvious path. Fieldwork is about coming together, about weathering the storms and celebrating the rushes. On Greenland, the challenge was enormous, so the highs were dizzying. Tracing meltwater over some tens of kilometres from the interior of the ice sheet using the same sensitive gas tracers employed to detect ocean currents travelling several thousand kilometres did not come without its difficulties. We worked feverishly under the Arctic sun, battling against icy winds that roared almost continuously off the ice, toiling to get our experiments to work. What was it Samuel Beckett said? 'Try again. Fail again. Fail better.'

On one early, doomed research trip at the end of 2009, I camped out with my Polish research assistant Greg (well, Grzegorz, but shortened to Greg) in a small tent at the margins of Russell Glacier, Leverett Glacier's smaller neighbour, where we were hammered by continuous autumn rain. Every day we trekked up onto the ice, wading across the swollen glacial river, shouldering heavy equipment while barely able to stand upright in the flow – right on the cusp of what felt safe and what really didn't. We lived from day to day, wet through, mired in mud and scrabbling around in the sands. On one disastrous occasion we lost most of our equipment stored beside the riverbank to a giant outburst flood as a lake trapped at the ice margin drained, tossing icebergs the size of cars downstream. It was so bad it was almost funny. A precious bottle of Jägermeister provided some cheer in the nights, and sometimes, when desperate, during the day. Nowadays I can't bring myself to drink this thick, pungent liquid – it dredges up memories of grime and failure. The gas tracers we injected into moulins on that trip never emerged at the ice edge, prompting much head scratching (and worse).

It took us another year to figure out how to make the gas-tracing method work. The problem was that the gas was so volatile that it went straight back to the atmosphere as the water plunged down the moulin. The solution? One hundred and fifty metres of garden hose to deliver it to below the water level of the moulin, and some muscle power to haul the hose back up again. ('You must have a very long garden,' the cashier in the hardware store had commented when we purchased the hose.) Once we'd nailed this, we rode a heady high of success; suddenly everything felt possible again. When we put a team next to a moulin forty kilometres from the edge of the ice, the gas tracers popped out of the front of the glacier in over twelve hours (roughly human walking pace), showing that the meltwater must have flowed via a fairly fast sub-ice river.[21]

However, when it came to injecting the tracer as far as sixty kilometres from the ice margin, something different happened. We hunkered down in a freezing tent on the banks of the tumultuous river draining Leverett Glacier, waiting and waiting against the continuous lament of the wind, stumbling out hourly to grab a tiny sample of water which we could later analyse in our field lab to see if there was any of the gas tracer in it. We had started to take shifts through the night, and had almost given up when the tracer started to appear. It had taken over eighty hours to travel sixty kilometres.[22] It was clear that the water's journey must have included some tortuous passages, before it reached the fast-flowing sub-ice river which popped out of the glacier front. This suggested that further inland, where the ice was very thick and less meltwater was being dumped down moulins, it was much harder for channels to form under all that pressure of moving ice. Instead, water must flow here partially at least by more sluggish pathways, seeping through soft sediments or weaving its way between interconnected cavities.

It was as though we had shone a light onto the bottom of the ice sheet for the first time. We enjoyed four summers of amazing discoveries, where each summer we would establish ourselves in our camp at the ice edge and push further and further inland from the ice-sheet margin, first on foot and then by helicopter. This heady period of endeavour and celebration took place in my mid-thirties, when I had a real zest for life. The occasional parties on Greenland were wild and carried on all night under the midnight sun. The remoteness of our camp meant that most of the time we had no alcohol, but when a newcomer arrived they came with fresh supplies, which always gave cause for celebration – and there was always the 96 per cent laboratory research-grade ethanol that Greg discovered went surprisingly well with tinned peaches and lemon juice! We'd spend the evening playing very loud music through tiny, crackly speakers, dancing on the grey sands, on one occasion descending into our own rendition of Led Zeppelin's 'Stairway to Heaven' with Marek on the air guitar, Greg on the fuel drum, a long-haired student from Minnesota on his prized set of panpipes, and me on the 'didgeridoo' (ten metres of plastic pipe used for mounting instruments in the river). All this, interspersed by search parties for moulins, injecting dye or gas into them, then days waiting at the ice margin for the evidence to emerge. I never slept more than four hours a night, but was kept going by an adrenaline-charged obsession to find out for the very first time how the Greenland Ice Sheet was plumbed.

As the newsworthy academic papers began to stack up with each new discovery, I started to wonder whether it might be possible to be promoted to professor. This had never been a goal for me. I had wandered into academia without much of a plan, to be honest, and when I looked at the generation of professors above me, there was never really anyone that I could identify with. The field of glaciology had always been a rufty-tufty, macho field, and senior women were few. But, with the funding and publications flowing, it started to feel like maybe it could be possible. As I saw my male contemporaries succeed, I started to wonder – why not me? Once I realized that such a lofty title might be within my grasp, I pushed myself even harder to achieve this goal. It took two goes. Crushingly, the first time I applied, the panel remarked 'not enough *Nature* or *Science* papers OR citations' – *Nature* and *Science* being academic journals where scientists battle to publish their most sensational (often controversial) findings in snappy form. The following year I submitted my application with little expectation of success. Much to my surprise, I was promoted to professor.

The research on Greenland was exciting because there was such a vast gaping hole in our knowledge. If the flow of water between the frozen underbelly of an ice sheet and its rocky bed is significant and greases the ice sheet's sole, then the big, overarching question which concerned all glaciologists was this: was more meltwater causing greater, faster slippage of Greenland's ice sheet into the oceans? At the same time as I was trying to trace the flow of the meltwater at its bed, Pete Nienow and his team had been busy scattering GPS devices at various locations over Leverett Glacier's surface to study the flow of its ice, and reaching the conclusion that no, it probably wasn't.[23] In fact, during 2012 – the biggest melt year on record at the time – the distance the glacier moved was no greater than in more normal melt years.

In essence, the ice sheet where it ended on land (rather than in the sea) had a clever way of self-regulating and preventing a doomsday cycle of more melt, more ice flow and more shrinkage.[24] Part of this self-regulation relates to the ice-walled river channels at the bottom of the ice sheet. Just like in Arolla, as more meltwater plunges down Greenland's moulins over a summer melt season, the frozen walls of these channels melt back, and the pressure of the water in them falls. Water naturally moves from high pressure to low pressure – it's the reason why concrete towers are used to store water in cities – and the low-pressure channels at the ice-sheet bed, just like the pipes at the base of the water tower, draw in water from the higher-pressure parts of the glacier bed, where it is perhaps flowing slowly via connected cavities or through soft sediments. This helps drain the glacier bed of water – but because the ice-sheet sole is no longer so well lubricated, it settles down onto its rocky bed. In the extreme melt season of 2012 the flow of the ice over the winter after the summer melt had shut off was diminished – because the bed had been drained of water by the vast pumping river at its bed. In fact, other scientists monitoring Greenland's ice in the lower melt zone have found many of its glaciers slowing down in recent decades, despite melt rates going up.[25] Basically, in places where the ice does not meet the sea, the ice sheet is looking after itself.

These fast-flowing sub-ice rivers didn't just help regulate the speed of glacier flow, they fed enormous amounts of meltwater into Greenland's seas, suspended in which were billions of tiny particles of that stuff called glacial flour, created by glaciers as they pulverized the rock beneath. The rivers appearing from underneath the ice sheet in summer flow like a frothy brown milk, and if you measure the amount of this glacial flour in the water, you can work out that roughly half a centimetre of rock is scoured from the ice-sheet bed every year.[26]

The discovery of all this fine flour pumping out of the ice sheet led us down a surprising new line of enquiry, of pertinence to Greenland's fjords and its vast surrounding seas. Soils are the skin of the Earth, supporting much of its life by allowing plants to grow. They do this by trapping water, providing nutrients and creating a soft layer within which roots can burrow, and so support their foliated canopy above. If we destroy our soils by overuse, we destroy the life-giving layer of our planet. Soil is basically a mix of rock matter and dead vegetation and, of course, a billion tiny bugs which work hard to decompose this dead matter. But soil's foundation material is rock, and it can take decades to break this down through weathering, in other words the gradual chemical and physical attack by the elements. Rock contains nutrients which support life by forming the building blocks of our cells; some plant nutrients, like phosphorus and potassium, are almost entirely supplied by its breakdown. Thus, you might think of the ice sheet as a gigantic nutrient factory, producing glacial flour, its ice rivers transporting all this fertile material to the oceans. But what happens to the flour and its nutritious load once it arrives in the sea?

Like so many things in nature, the answer to this question is not nearly as simple as you might first think. The rather complicating thing about glacial flour is that although it contains nutrients, unfortunately it makes rivers so cloudy that light can't easily pass through. After emerging from beneath the ice sheet, the murky glacial meltwater gushes into a dense network of fjords, the icy equivalent of estuaries. These were carved deep during past glacial periods, then mercilessly drowned as the great ice sheets melted and sea levels rose. In Greenland's landfast sectors where outlet glaciers don't reach the ocean, largely in the south-west and north-east, rivers dump their cloudy load into fjords at their surface. Satellite photos reveal this in the form of murky brown plumes of water spreading out from the land – and more melt basically means more murk.[27]

Light is the driving force behind the complex food webs of the sea. It fuels the growth of tiny plant-like organisms called phytoplankton, from the Greek *phyton* (plant) and *planktos* (drifting). These minuscule plants use the sun's energy to trap carbon dioxide from the atmosphere and to grow by photosynthesis; they then act as food for larger creatures which can't use solar energy, initiating a food chain whereby big things eat little things until you arrive at fish, seals, walrus and so on. One might say that phytoplankton are the food-makers of the ocean, a bit like crops are on land for humans. But in Greenland's fjords headed by glaciers ending on land, cloudy melt plumes cut out light and stifle the growth of these tiny plants. The fresh glacial melt-waters also tend to float on top of the fjord, trapping salty, nutrient-rich ocean waters beneath them, out of reach of the phytoplankton. As a result, fishermen catch little here[28] – the glacial meltwater is starving the fjords of life.

Yet if we take a trip to Greenland's fjords headed by tide-water glaciers, it's a different story. Here the glaciers' elongated tongues float gracefully on the water surface, calving giant icebergs as they become unstable in deeper water. Curiously, when the sub-ice rivers blast out of the glacier snout, they're submerged beneath some hundreds of metres of salty fjord water – making the margins of tidewater glaciers a bit like a giant jacuzzi. When you sit in a jacuzzi, jets of hot water spurt out from the sides of the tub below, bringing heat and bubbles to the surface; whereas with these Greenlandic 'jacuzzis', the jets are a mix of cold glacier melt and slightly warmer ocean water from the deep. These jets melt the tidewater glacier ice fronts as they rise to the surface; they've been a major cause of the recent rapid retreat of these outlet glaciers, where both ocean warming and more melt from the sub-ice rivers have strengthened the jacuzzi melting effect.[29] However, as the warm jets rise up from the depths, they also bring up the nutrients the phytoplankton crave – most importantly of all, nitrogen.

Phytoplankton need a balanced diet to grow and reproduce. Carbon comes top of the list – they can get that from the carbon dioxide in the air – then nitrogen, phosphorus, followed by smaller amounts of micronutrients like iron, zinc and so on. Depending on their habitat, phytoplankton may run out of one or two nutrients before the others. In these cases, we say that the phytoplankton are 'limited by' the nutrients they have run out of. To use a basic human analogy, if all you ate was plain rice, you would lack protein. However much rice you consumed, this would not help you build more proteins in your body, and put on muscle – you would become malnourished and weak. In the fjords of Greenland, the big problem nutrient for phytoplankton is nitrogen. Ocean waters which seep into fjords at depth are packed with nitrogen,[30] but the fresh glacial meltwaters which flood the surface layers are sadly packed with phosphorus, silicon and iron, rather than nitrogen.

So with tidewater glaciers, the extra nitrogen delivered up from the deep via 'jacuzzis' nourishes the phytoplankton at the fjord surface,[31] while the light-quenching particles often drop out of the plumes before they can get to the surface. The benefit to these tiny plants cascades all the way up the food chain to fish, which bring in more than 90 per cent of Greenland's export income.[32] Halibut, accounting for nearly half the territory's fisheries income, often lurk in the entrances to fjords[33] – a correlation has been spotted between the water emerging from tidewater glaciers via jacuzzi-style rivers and the size of the catch.[34] As long as glaciers continue to float in the ocean in Greenland, it's more likely that halibut will be on the menu.

We're not quite done with the glacial flour. What happens to it now? Even if it has a detrimental effect in the fjords, does it have a useful impact anywhere? The answer is that in Greenland we still don't know for sure. A few clues exist, though, that some of these tiny particles journey out into the seas off Greenland, borne by ocean currents.[35] Here they don't cut out light, but they still contain tasty nutrients – especially silicon, phosphorus and rather a lot of the micronutrient iron, which is released from pyrite (fool's gold) and other iron-containing minerals entombed in the rockbeds of glaciers.[36]

Amazingly, pictures taken from space are able to detect the pigment chlorophyll that is found in all living plants, including phytoplankton, which changes the colour of the ocean surface. It seems that this green pigment becomes more abundant in the Labrador Sea, offshore from western Greenland, in midsummer – which is also the time when the ice sheet is melting fastest. The fresh glacier meltwater, carried offshore by currents, seems to be bringing nutrition to the phytoplankton and helping them grow. A similar phenomenon has been observed in the Southern Ocean around Antarctica. Here, phytoplankton don't have enough iron, mainly because this is typically supplied by dust from deserts, and this ocean is extremely distant from places like the Sahara. However, icebergs drift out from the Antarctic continent and dump iron from dirt entombed in their frozen bodies as they slowly melt. This stimulates the growth of phytoplankton, boosting chlorophyll and changing the colour of the ocean.[37] Essentially, it seems that the ice sheets are fertilizing our seas in both the Arctic and Antarctic.

In Greenland, however, tidewater glaciers are fast shrinking back onto the land as the oceans around them warm, with potentially disastrous effects on local fish and other wildlife, as the phytoplankton – food-makers of the fjord – are stifled by the murky waters at the surface. This is less likely to occur in the north-east of Greenland, as the bottoms of these glacier tongues are well below sea level – more likely that as they retreat back, the sea waters will simply flood into the hole created behind them and they will keep their floating tongues – but elsewhere on the island it is perfectly possible.[38] Here, fjords may become decreasingly productive, as the tidewater 'jacuzzis' are slowly turned off[39] and phytoplankton are cut off from their deep marine nitrogen supply.

These are big changes, which will affect those who depend on the seas around Greenland for their subsistence. Both the glaciers and the sea ice have long been important resources to the local Inuit, who arrived in north-west Greenland via the Canadian Arctic, Hudson Bay and Labrador;[40] they now inhabit small coastal communities around the fringes of the ice.[41] The winter sea ice offers a platform for hunting, and together with icebergs from tidewater glaciers, provides important resting places for seals and other mammals. Now that the ice is thinning and present for shorter periods, the question is, what impact will this have on the Inuit?

These people have already endured some very dramatic climatic shifts during their long history. They have flexible, mobile livelihoods which have fostered in them a deep-rooted ability not just to adapt to but also to anticipate change.[42] This is well exemplified by the 700-strong community inhabiting the remote far north-west of Greenland, mostly in the town of Qaanaaq, who are known as the Inughuit but often referred to as the Thule Inuit because they originally arrived from Canada

1. My emotional return to the Haut Glacier d'Arolla in July 2018 – its sad snout sitting like a ghost in an empty valley, walled by the Bouquetins Ridge.

2. My place of awakening after a long, cold night on cardboard during the summer of 1992, facing the tumbling Bas Glacier d'Arolla ice fall.

3. Tall, lanky (or in his words 'long-limbed, lithe and elegant like a pedigree race horse') Pete Nienow waiting for the arrival of rhodamine dye in the proglacial river at Arolla.

4. On the Haut Glacier d'Arolla, water plummets down a moulin to a dark, hidden underworld that teems with microbial life.

5. Cold, barren and patrolled by bears: Finsterwalderbreen gleams in the Arctic sun.

6. At Finsterwalderbreen, the channel that inspired me in spring, and caused me a fractured kneecap during a freak flood in summer.

7. The air temperature is −20 degrees Celcius, yet water gurgles beneath Finsterwalderbreen's winter naled ice – where has it come from?

8. The unnerving prospect of sampling Finsterwalderbreen's violent chocolate upwelling.

9. Pondering the treachery and mystery of a crevass slicing through the surface of the Greenland Ice Sheet.

10. Happy to see the sun after a long, lonely night in Greenland spent sampling Leverett Glacier's river for elusive tracers.

11. Heart in my mouth – we attempt our first major dye trace to a moulin 15km from Leverett Glacier's margin.

12. The myth that is 'glacier blue'...

and settled in the region of Thule in Greenland.[43] Their lives are inextricably bound to the riches of the sea, hunting for seal and narwhal. They arrived here in about AD 1100, a clement time known as the Medieval Warm Period, across the slim land bridge that existed at the time between Arctic Canada and Greenland, from which they spread south.

But in the fifteenth century, the cold period called the Little Ice Age took hold in Greenland and elsewhere around the world. At this time, the Inughuit became cut off from any settlements in the south through the advance of glaciers and there being simply too much ice. It was not until 1818, when Captain John Ross reached the region in search of the Northwest Passage, that these people were 're-discovered'. He wrote that when he met the Inughuit hunters on the ice they had 'believed themselves to be the only inhabitants of the universe, and that all the rest of the world was a mass of ice'.[44] Climate change has always been a fact of life for these people.

With the recent warming, they face the exact opposite situation – sea ice being on the way out – but things are more complex for them now, because they are connected into a global economy and political interests. On the upside, halibut have migrated northwards, providing a new income stream,[45] while the opening up of oil, gas and mineral reserves has been heralded as an 'opportunity' for commercial exploitation.[46] On the downside, many hunters have had to kill their dogs due to the lack of sea-ice cover, the crossing of which required sled teams. What will happen is unknown, but it will not be without strife.

I pondered these issues, among others, during the long days at our tiny camp in front of Leverett Glacier – a throng of sun-bleached tents nestled like a pod of whales between the humps of bouldery moraines, once beneath the glacier, now exposed as a disorderly collection of rubble. This was the dustiest, windiest place I had ever been; the loose grains of sand and silt managed to worm their way into every item of clothing, hair, nose, ears and even more worrying places. Gear like rucksacks and tents bore the scars of these dreaded particles for years to come. The minuscule grains clogged the plastic grooves of the zips of our tents until they failed; the most effective remedy was to sew Velcro down the sides of the doors, which took for ever through the thick canvas. Being a habitual insomniac, sleep never comes fast for me in these places – continuous summer daylight, the whirling wind and the roar of the river always occupied my mind. I both looked forward to and dreaded fieldwork in Greenland, eager to be back in the wilds and on a big mission, but nervous because I hated the hollow feeling caused by constant sleep deprivation.

The exhaustion was softened somewhat by an astonishing backdrop of ice cliffs, energetic rivers, serene melt ponds and herds of musk ox, which would boldly wade in the shallows of Leverett Glacier's wild river and wander through our camp. These are curious animals, introduced to south-west Greenland from the northern part of the country in the 1960s, their silken locks draped like brown skirts over their stumpy bodies, a bit like a North American bison. Their long hair has special qualities – it boasts an outer layer which sloughs off moisture and an inner layer that is much softer and extremely warm and light.[47] Musk ox have always been prized for their wool, hides and meat by the Inuit here, and once came close to extinction in western Greenland, necessitating their reintroduction from the east. One of their most fascinating traits, which I observed several times, was that when threatened by wolves or people, they gathered in a defensive circle, facing outwards with their horns lowered menacingly as protection. Yet as soon as they began to run, they reminded me of giant guinea pigs – hooves scrabbling at speed to propel their hefty shaggy bodies forwards. They always brought a smile to my face. They are also the cause of my brief lapse from vegetarianism while I was in Greenland – after twenty years of not eating meat, I became so worn down by the hard physical work that I succumbed one day to a 'Musk Ox Burger' in Kangerlussuaq airport, and then another and then another. (I have now thankfully mostly managed to 'lapse back'.)

High above our camp was a rounded hill, its hard bedrock surface polished by the ice sheet in cooler times. In a grassy hollow on the flank of the hill lay a tiny lake, its dark glossy surface tranquil above the wildness of the glacial river which gushed past below on its way to the sea. This was a place to which I often went to escape camp life, to clear my head. I was almost always joined by a raucous pair of geese in summertime, their honking calls rebounding around the yawning valley. From this vantage point I could see all the way to the fjord by the town of Kangerlussuaq, the river of Leverett Glacier joining forces with several others to spread chaotically across the proglacial plain (or sandur), glistening in the sun like a tangle of silver threads. In one direction all was grey and white, in the other everything was green – the Green Land.

What was it that spelt such different fates for the Inuit, who survive to this day, and the Norsemen of Greenland, who are long gone? Both must have possessed a resilient spirit to establish themselves in this harsh land, and a strong sense of community to weather the challenges they faced. The demise of the Norsemen has always been cloaked in mystery – they grew to populations of several thousand in two major settlements in the south of Greenland, only to collapse five hundred years later, leaving behind but the ruins of former stone houses and farmsteads in the far south and west of Greenland. Like the Inuit, they arrived in Greenland during the Medieval Warm Period, between AD 900 and 1400. Between 1350 and 1450, however, the climate in southern Greenland became colder and stormier as the Little Ice Age started to encroach – the Norse disappeared around AD 1450.[48] This convergence in timing might suggest the inability of the Norse to adapt their farming practices in the face of inclement conditions. (Unlike the Inuit, for whom the greatest riches lay in the sea.)[49] But recent evidence suggests that the Norse did in fact adapt quite well, and that a significant part of their livelihoods involved harvesting the seas around Greenland, in particular for walrus, whose tusks[50] were a valuable source of ivory that was used to make luxury products in medieval Europe.[51] In fact, DNA taken from ivory artefacts from the time of the Norse shows that they probably had a complete monopoly on the European walrus ivory trade. Yet the success of the Norse, it seems, may have been their downfall – an over-reliance on walrus in changing times.

The story of these hardy northerners ought to provide some lessons.[52] They had invested heavily in fixed settlements rather than being migratory; they had relied strongly on a single commodity, walrus tusk, which it is thought they over-exploited; and their relationship with the Inuit, who had more diverse hunting strategies, was far from collaborative.[53] This was all set against a backdrop of colder times, the decimation of populations in Europe caused by the Black Death, and plummeting demand for walrus ivory as elephant ivory flooded the market. No one has found evidence of walrus tusk being exported from Greenland to Europe after about 1327,[54] and perhaps herein lies a clue to why the Norse died out in Greenland. Sadly, all theories rely on the preservation of artefacts, many of which have been buried in the permafrost and are now being lost fast as the ground thaws and organic remains decay.[55] Perhaps the truth will be lost with these precious remains.

But whatever it was that brought about the demise of these Nordic farming-hunting folk in southern Greenland, their story is closer to home than we might judge at first glance. Our climate is changing dramatically, at a time where the world is increasingly connected, where markets are truly global and where pandemics are catalyzed by air travel. The Norsemen exemplify both resilience and vulnerability in the face of a changing climate;[56] maybe they would have survived if they'd forged a closer relationship with the Inuit, if they'd been able to migrate to keep the harsh climate at bay, and if they'd not relied too much on a single market product.[57] Perhaps some clues to our future lie in their fate.

# 4. *Life at the Extremes*

*Antarctica*

Days had passed, and nights too, almost indistinguishable from one another under the relentless glare of the sun. He'd travelled far from the sea; while out fishing one day, he'd got separated from his penguin family and had emerged onto land suddenly confused about which direction to waddle. Normally he'd use the sun to guide him, but dense cloud cover made this tricky. So he'd started to walk away from the sea, to cross the headland – only it wasn't a headland – and the sterile, white surface yielded no clues as to the true path, so he'd just kept going. He was unbearably hungry. Then, one day, on the horizon, he'd glimpsed a cluster of jaunty domes – canary yellow and orange against the whiteness.

He was an Adélie penguin, the smallest and one of the most common species of penguins along the Antarctic coast, which are capable of diving for up to six minutes underwater and down to depths of one hundred and fifty metres in search of fish.[1] These comical-looking birds were encountered by a French expedition in the nineteenth century, and named after Adélie Land, a portion of the Antarctic mainland which the explorer Jules Dumont D'Urville had in turn named after his wife Adèle.[2] Apart from my human companions, of course, this little penguin was the first living creature I had set eyes on in Antarctica. One day he'd stumbled into our camp lost, his navigational senses scrambled. We were about one hundred kilometres from the sea – not to mention a supply of fresh fish – having travelled by helicopter for an hour or so from the Antarctic base of McMurdo the day before.

You're strictly prohibited from interfering with wildlife in the McMurdo Dry Valleys; no feeding of animals and birds is allowed, and you must let nature take its course. So there was nothing to be done with our newly arrived black-and-white friend. In the beginning he entertained us with his antics, blundering around, wings aflap, getting caught up in the fly sheet of the mess tent, entangled in the guy ropes. He made us laugh – there's nothing elegant about a penguin on dry land. Our meeting with the little Adélie was bitter-sweet – sweet because that is indeed what he was, bitter because we could not help him.

Antarctica truly is Earth's last great wilderness, the one place where humankind has so far failed to put down roots. It is so far from anything we can imagine or feel, a mysterious, inhospitable white void in a world atlas, its name giving nothing away – meaning simply 'opposite the Arctic'. Yet this far-off continent is about the size of Canada and is dominated – except for about 2 per cent of its area – by our greatest ice sheet, separated into east and west lobes by the twisting spine of the Transantarctic Mountains which connect the Ross Sea on one side and the Weddell Sea on the other.

It is surrounded by the treacherous Southern Ocean, its waters kept cool by the intense Antarctic Circumpolar Current (ACC), which swirls clockwise around the continent, fending off warm waters from the north and helping preserve the ice sheet. At its northerly boundary, the ACC dives beneath the more tepid waters of the sub-Antarctic, and the mixing of the two brings nutrients to the surface, sustaining life and particularly the prized Antarctic krill. Krill is a semi-transparent, shrimp-like crustacean, what we call a type of zooplankton, that drifts around the ocean following currents. It's one of a few types of Southern Ocean marine life that can feed directly on something as small as a phytoplankton, which it does by filtering them from the ocean water using its tiny feathery legs. Many larger forms of life eat krill, and it sustains valuable fisheries for nations like Norway and Japan – without the krill the food webs of the Southern Ocean would collapse. Thus, despite its roughness, the Southern Ocean teems with marine life – whales, albatross, penguins, seals, krill and many species of fish flourish here.

Antarctica has always offered a benchmark, by way of its extremity and remoteness, against which to determine one's purpose in this world; throughout the intensive era of polar exploration and into the present, it has fulfilled a deep-seated human desire to venture into the unknown. Many pilgrimages have been made to this featureless, colourless land. Everyone who has gained an acquaintance with Antarctica has a word for its vast polar desert – awe-inspiring, blank, hostile, isolated, serene, stark, sterile, striking . . . unforgiving. For me, it embodied desolation – for my visit there in 2010 coincided with one of the most desolate periods of my life.

My expeditions to Greenland had been wild, euphoric and charged with ambition and hope. Yet it was during this period, too, that my mother had been diagnosed with advanced breast cancer, which had progressed through much of her body. She had responded brilliantly well to the chemotherapy drugs, in part, I always felt, through having a spirit that was more positive than anyone I knew. Two years down the line, though, the drugs were no longer working quite as well and causing severe side effects. One spring during this bleak time the opportunity cropped up to visit Antarctica. Totally out of the blue, I had received an email from a scientist from New Zealand, John Orwin, who'd been working in Antarctica with Martin Sharp, my glaciology professor from the Arolla days. His small team from Dunedin in the chilly south of New Zealand were on the hunt for someone who understood both glaciers and water chemistry and who might be prepared to join them on a trip to the McMurdo Dry Valleys to collect some data to support a forthcoming research grant proposal. As always, the call of the ice was too strong to resist.

It felt desolate from the start. On Christmas Day I sat alone in a plush apartment in Christchurch, New Zealand, carefully unwrapping the brightly coloured presents which had been hidden in my backpack by my family. The pain engendered by the distance and separation was excruciating. What if my mother were no longer alive by the time I got back? How would I know? What on Earth was I doing on the other side of the world, in Antarctica of all places? Why couldn't I just be like a 'normal' person, who would surely stick around at a time of family crisis? Where did my constant restlessness come from? The next day I was strapped bolt upright into a military plane, a thunderous roar in my ears, bound for the southern continent. There was no going back now.

Stormy conditions tend to prevent helicopters from taking off, and so we spent the first few days grounded at the New Zealand research station, Scott Base on Ross Island, named after Captain Robert Falcon Scott, who had departed from the shores of the nearby Ross Ice Shelf during his fateful Terra Nova Expedition of 1910 to 1913. (Not only had the British party been beaten to the South Pole by Roald Amundsen's Norwegian team, but they had all perished on the return journey from the South Pole.) I have always found research stations strange, claustrophobic places. The atmosphere can be heavy with extreme politeness, as people who don't really know each other (and probably never will) attempt to make light conversation over meals; drowsy from the suffocating warmth of the supercharged heating systems and the insomnia that comes from sharing bunkrooms with snoring strangers. In the event, our confinement at Scott Base could have been a lot worse. Its cheery crew were used to making their own entertainment, and while I was there the twenty or so permanent station staff staged an entire (mock) wedding where the burly carpenter married the equally burly cook, complete with wedding ceremony, dinner, speeches and a reception party got up in drag. I'd also managed to purloin some skinny cross-country skis – I had no idea how to use them, but my Canadian colleague, Ashley Dubnick, who I already knew from my Greenland exploits, kindly gave me an impromptu lesson on the Ross Ice Shelf. She glided like a swan over the glassy ice-shelf surface, rhythmically loping from one foot to the next in an elegant, fluid motion. I, on the other hand, resembled Bambi – a few strides followed by a fall to the ground, having lost my balance for the umpteenth time.

Finally, after several days of limbo, we flew by helicopter to our destination, the McMurdo Dry Valleys, the largest expanse of ice-free land on the Antarctic continent. Nestled between the Transantarctic Mountains and the Ross Sea, these barren valleys are starved of both moisture and warmth – even in summer, the air temperature mostly bumbles along below zero degrees Celsius. A peculiar collection of glaciers flows down into them from the mountains. These are 'cold-based' glaciers, because many of them are frozen to their beds, and they flow ever so slowly, moved by the minuscule dislocation and stretching of ice crystals under the pressure of an enormous frozen mass. Their swollen, white lobes feed glacier tongues which protrude like giant alabaster slow worms from the valleys, exposing ice cliffs which tower imposingly over the sandy plains below. These near-vertical cliffs are impossible to assail without the aid of a sharp pair of crampons, an ice axe and some courage. Their steepness reflects the fact that there is little melting of ice here – so the snouts of glaciers in the Dry Valleys do not develop the typical gently sloping surfaces of glaciers in, say, the Alps. The sub-freezing air temperatures throughout the year in this part of the Antarctic have helped the glaciers maintain relative stability compared to others elsewhere; the climate-warming time bomb has yet to fully hit them.[3]

My time in the Dry Valleys amounted to six weeks, yet it felt like six months. There were none of the highs of Greenland, no crazy parties, no monumental scientific breakthroughs to elevate the spirits. Just desolation and drudgery. A small team of just three scientists – Ashley, John Orwin and me; we were entirely alone in the deep field. With no phone, no email, no radio, I was tortured by worry about my mother back home and felt an overwhelming sense of disconnection. Unlike my companions, who abstained from alcohol for the entire trip, I had to partake of a dram or two of single malt every night to bring relief from my spiralling thoughts. It was a treasured ritual – I'd carefully decant a portion of thick, golden Islay malt into a 28-millilitre glass vial, originally intended for collecting water samples. It became the highlight of my day, which sounds dismal given that I was in Antarctica, a place most people can only dream of visiting.

We established our small camp on the shores of Lake Colleen, in Garwood Valley at the far southern end of the Dry Valleys, which borders the McMurdo Ice Shelf and the vast Ross Sea. I'd expected these valleys to be arid, but still was amazed by what this meant in reality. Of course, like everywhere in Antarctica, the snow tumbled from the clouds – yet, within an hour of touching the ground, it had vanished back up to the skies. I could scarcely believe my eyes. It couldn't have melted – the air temperature was far too cold. Instead, the air's extreme dryness meant that the solid snow had transformed, as if by magic, straight back to water vapour – a process called sublimation, when a solid goes to a gas – thus avoiding the liquid phase that would occur if you heated a chunk of ice in a pan on the stove, when it would turn first to liquid and then to steam or 'water vapour' (a gas state). This sublimation of snow is the main means by which glaciers in the McMurdo Dry Valleys lose mass, although they do also melt a bit in summer.[4]

Due to the extreme cold and aridity, this landscape was littered with the mummified remains of seals, penguins and other benighted creatures which had wandered there by mistake. Their leathery remains can be preserved for as much as thousands of years[5] in the frigid air, which protects them against breakdown by bacteria and other microbes. This is not to say that microbes don't exist in the Dry Valleys, but they are not numerous compared to, say, a tropical soil, and they work slowly under such conditions. It was Captain Scott who initially discovered these valleys – referring to Taylor Valley as the 'Valley of the Dead'.[6] Curiously, it was my first mental image of these valleys as a graveyard where most forms of life struggled to survive that provided my reason to visit. Drawing on my earlier discoveries about microbial life and how it survives deep beneath glaciers, I wanted to know how life thrived on the glaciers here, in one of the most challenging places on Earth, not to mention what this might tell us about the habitability of the huge ice sheet that lay silently next door, and even perhaps of other icy planets and moons.

Despite the eerie backdrop of corpses strewn across a polar wasteland, our small gathering of tents was the most luxurious field camp I'd ever set foot in, probably because it wasn't run by a Brit – we tend to do things the hard way. The Kiwis, who were running the logistics for our trip, had helicoptered in a chest freezer powered by a generator, which was brimming with pre-packaged bags of meat and tasty sauces that cohabited with our ice samples. If you do research in Antarctica, you normally carry it out in association with a national logistics operation – whether it's the Brits, Americans, New Zealanders or one of forty or so other countries with a presence there – who will supply the right gear and make sure things proceed safely. This was a far cry from my work on Greenland, where my small team of hardy glaciologists doubled up as the logistics operation, managing our own transport, camp set-up, kit, power, food and daily regime. In Antarctica, I didn't have to worry about any of that.

Our mess tent, aptly named a 'Weatherhaven', was fit to survive a tornado – it resembled a giant yellow banana that fanned out like a concertina, its shell stretched taut across a series of hooped poles. I'd had mixed experiences with this tent from the start, when Ashley and I tried to put it up after being dropped by helicopter during a very short weather window (we had had to leave John behind because of the huge amount of equipment that needed to be stowed away – he joined us later). We'd never seen anything quite like this tent, and were initially quite puzzled, but we'd managed to figure it out and assemble it into its bizarrely shaped body, with all the poles correctly installed. We were inside, admiring our creation and marvelling at its all-round yellowness when a powerful squall swept across the proglacial sands. Without warning we found ourselves violently tossed back and forth as the tent was swept up and propelled down the rocky incline towards the lake, as we desperately rolled around inside, trying to stay upright, scrabbling frantically like hamsters in a turning wheel. We'd forgotten to peg it to the ground. Luckily, we ground to a halt when the wind abated, just before starting a wild slide across the glassy lake ice.

Our personal tents were thankfully more simple affairs. Mine became my safe place – a refuge from storms of the weather and the mind. The New Zealanders had provided not one but three sleeping bags – a fleece liner, an inner and outer down bag – all of which were laid out upon a thick insulated air mattress covered with a woollen fleece pelt. Strange to say, I have never slept so well as I did down in the Dry Valleys. Despite my internal turmoil during the day, when night fell my mind and body seemed to settle down into a slower rhythm, and I felt cocooned in my tiny tent, a warm, tingling memory of whisky in the back of my throat.

My mission was to find life in this desolate landscape, but the one basic thing that life needs to survive is . . . water. And here I was in a place called the Dry Valleys. Cells die without water. Humans can survive just a week or so without it; we're made up of cells, after all, and more than half our body weight is water. But for a single-cell microbe, no water would be a disaster. So my first job in the Dry Valleys was to seek out a source of water. At first glance, this might seem unlikely in a place where air temperatures were already some way below freezing – and yet the melting of glaciers in the Dry Valleys still occurs. The sun here is strong, and in summer it casts its most powerful rays – often in the form of short-wave radiation – onto ice surfaces round the clock. Dark particles of dirt entombed in the surface ice of the glacier absorb these rays and start to warm up, just enough to thaw the ice around and, especially, beneath. The dirt slowly melts its way down deeper into the ice, leaving a molten liquid capsule just above, and over time something called a 'cryoconite hole' forms. These are literally tubes of water, with dirt (called 'cryoconite') at the bottom, surrounded by circular ice walls.

If you stand on a melting glacier almost anywhere in the world and peer down at its surface, you'll see hundreds of these holes sunk into the ice, almost like it has been perforated by a giant hole puncher, tiny amounts of dirt winking at you from the bottom. But in the Dry Valleys, the cold air often keeps cryoconite holes frozen at their surface, so they have a thick ice lid, a bit like a frozen jam jar.[7] This, I discovered, makes them extremely tricky to spot. What is fascinating is that the water capsule trapped beneath the ice lid usually stays liquid in summer because the dark sediment at the bottom of the hole is always being heated by the powerful rays of the sun and the frozen ice lid provides a little insulation from the cold.

Thus on my first trip up onto the nearby Joyce Glacier, a two-hour trudge away, over sands, rock and then ice, I was puzzled – where were all the cryoconite holes? The glacier descends from the Royal Society Range mountains, which cradle its flat, dazzling white trunk before it terminates abruptly in striped ice cliffs, displaying past layers of dirty and white ice in their chiselled vertical walls, and is frozen to the rock beneath it, so there are no rivers, swamps or cavities concealed beneath the ice. I did, as it happens, find a small number of cryoconite holes on the ice surface – but the discovery that captivated us was its collection of enormous ice-lidded melt ponds, some metres across, laden with sediment at the bottom and topped by a solid block of ice about as tall as a baked-bean tin. Similar ponds had been found not long before on Canada Glacier, close by in the Dry Valleys, acquiring the term 'cryolakes'.[8] They seemed to be clustered at the edges of the glaciers and were sometimes interconnected by tiny watery channels also with ice roofs – creating a sub-ice plumbing system, but this time on the glacier surface. These were the final stopping point for meltwater produced on the glacier before it poured over the ice cliffs and into rivers and lakes. The cryolakes were pretty cool – you could slide on your feet across their smooth frozen surfaces, and gaze down through their icy lids into a watery underworld – an oasis in a polar desert.

Until about a decade before my trip to Antarctica, it was generally believed that glaciers were devoid of life – simply sterile wastelands that could be largely ignored when thinking about Earth's natural history. Following the astonishing discovery of microbial life beneath the Haut Glacier d'Arolla, though, there was a big push by researchers to find out whether glaciers elsewhere were furnished with such tiny life forms. If life in glaciers was ubiquitous, it would mean that the ice-encrusted poles and high mountains could no longer be deleted from the complex web of stores and flows of carbon on Earth (called 'the carbon cycle') and when considering its biodiversity. It was also possible that these hardy life forms boasted useful adaptations to the extremes of cold, to harsh radiation on glacier surfaces or to heavy metals like mercury released from sediments at glacier beds – perhaps they contained genes that coded for the manufacture of sunscreen pigments, or which converted metals to less toxic forms?

So how might I go about finding signs of life – the type far too small to see with the naked eye – in a cryolake covered in thick ice? Luckily, I'd recently started working with a bunch of brilliant engineers from the UK's National Oceanography Centre in Southampton, who were experts in developing and testing technologies around this very area – the kind of instruments you might use, say, to detect life in remote parts of our oceans or even on other planets. One of them was a young engineer called Matt Mowlem, who was well known for trail-blazing new ways to detect nutrients in the ocean using minuscule flat, plastic chips the size of a matchbox; in essence, Matt and his team took bulky bench-top chemical instruments of the type I had in my lab and downsized all their components onto a chip – creating sensors which were about the size of a 1.5-litre water bottle, called 'Lab-on-Chips'. Everything about Matt was pragmatic – if he ended up in a hotel room where the fan or the lights didn't work, he'd fix them. If he had too many pigeons frequenting his garden in Southampton's suburbs, he'd pop them with his air rifle and eat them for dinner. I always admired his ingenuity.

Matt and I had recently received government funding to take technologies, including his Lab-on-Chips, developed for less extreme environments, and test them out on glaciers. One invention that piqued our interest was a sensor which measured the amount of oxygen dissolved in water – this was a good starting place for the search for life in our cryolakes. Oxygen is a good marker for life in these sealed melt capsules, because if it is found at lower levels than in the atmosphere, this would suggest consumption by living things; whereas higher levels might indicate production by the process of photosynthesis. Scientists had already spotted algae in the smaller cryoconite holes on Taylor Glacier nearby and in the ice-covered lakes of glacier forefields in the Dry Valleys, which gave me some hope of success.[9] Even in these icy places, though, it wasn't clear how life could be supported under such extreme conditions. If I could get beneath the ice roof of the cryolakes, I could zap their waters every few minutes with my instruments during the peak summer melt period to see how much oxygen they contained.

It took Ashley and me almost a week to lug the necessary bulky items of equipment up onto the glacier – kit that included dataloggers (a bit like simple computers to record data in time), a variety of instruments and their spaghetti of electrical cables, metal poles, miscellaneous tools, solar panels, batteries and, most importantly, duct tape. (The solution to just about every problem under the sun, from holes in down jackets to flapping cables and crates bulging at the seams – glaciologists never go anywhere without duct tape.) I'd already worked with Ashley in Greenland; she'd been invaluable when we were trying to figure out how to inject our gas tracers down moulins at Leverett Glacier, and vast lengths of garden hose had needed to be hauled up onto the ice sheet on foot. She was tough.

Every morning, when I awoke to the faint, orange glow of my tent canvas above, I'd greet the new day with a feeling of peace, only to be swiftly banished by stomach-lurching panic as I remembered my mother's predicament back home and the distance that separated us. Breakfast was an ordeal to be weathered – the need to engage in social chit-chat over muesli, while battling a heavy blanket of internal despair. After that, rucksacks were packed and we'd set off for Joyce Glacier. The first part of the journey, across the frozen surface of the lake next to our camp, was easy until the ice gave way in a rotten section and one's leg disappeared down a wet, icy hole. Then there was the trudge over the sands, between the dune-like moraines in front of the glacier, which shifted beneath our feet as we struggled with our weighty loads. We always fell into a line as we plodded, one by one, like beasts of burden crossing a desert. This part of the journey was emotionally gruelling for me, and I was always thankful to be separate from my companions, because often tears would be streaming down my face as the anxiety worked its way out of me. As in Svalbard, music was my saviour – this time delivered by an iPod rather than my treasured Sony Walkman of times gone by. A blast of energetic guitars and rousing drums helped me reboot myself, and rescued me from despair.

The final slog up and over the ice cliffs at the edge of the glacier was the most excruciating part, especially with twenty kilos of random kit strapped awkwardly to the perimeter of one's rucksack, inside which was crammed an assortment of emergency clothes, food and water, all packed around various tools, batteries and technical whatnots. The narrow gulley up the ice cliffs was crampons and ice-axe territory. I've always been pretty agile on my feet and have a good sense of balance, acquired through years of swinging from the rock cliffs of the Avon Gorge in Bristol. Although far from vertical and not technically challenging, the thought of slipping and toppling backwards during this ascent always added to the sense of jeopardy. I would feel the familiar sharp bite of my rucksack straps over my shoulders and shooting pains from trapped nerves blasting up my neck as I stooped over, willing my leaden legs upwards over the ice lip. These pains are fairly common, occurring when you have far too much weight in your ruck-sack for your body to carry effectively. Gritting one's teeth is helpful in these situations, I find, as it somehow leads to the gritting of one's mind. You simply get used to the pain – it just becomes another sensation, like heat or cold or the feeling of movement in the air. You notice it, have a short conversation with it, then crack on.

I may not look like a rugged outdoors type – weighing just eight and a half stone and standing five foot seven inches tall – but I suppose I'm a lot tougher than I appear. Over the years, I've weathered so many baffled facial expressions and blinkered comments – ranging from 'you don't look like a glaciologist' to 'how do you cope with the cold, there's not a lot of fat on you?' – that I've learnt to smile politely and move on. Don't give it any energy. Yet there are certainly aspects of life in the field which, as a woman, are definitely more challenging.

One thing I often notice young female students panicking about when bound for the field is – how do you pee on a glacier? On a blank, featureless expanse of ice and rock, there are few places to conceal oneself. My original ploy, following pure logic, was not to drink *anything*. In the early days, when I was often the only woman on an expedition, this is how I saved myself from embarrassment. Twelve hours would pass, during which I'd miraculously manage to command authority over my bladder to simply – contain all. But this path of self-denial was clearly unsustainable (indeed, potentially harmful), and so it was sometimes necessary to disappear furtively behind a boulder, down a snow pit or occasionally simply to go in the wide open space, following a muttered request to colleagues to avert their eyes.

In the Dry Valleys, the problem of dealing with this simple bodily function became something else again. You see, in this part of the world you can't pee freely. Quite literally all urine, indeed all waste, has to be collected up and helicoptered out at the end of the season – this is integral to the 'do no harm' code of conduct binding scientific research in these valleys. So at the start of this particular field campaign, the Kiwis (an enlightened people) had provided every female team member with a 'Shewee' – in essence, a plastic tube to be placed where needed, when needed – and a plastic bottle to collect the 'proceeds'. This was a total revelation. For the first time in my career I had some of the same advantages of a man. I no longer needed to find that rock or snow hole, or have that awkward conversation with my colleagues. I could just saunter off into the distance – how liberating!

The journey from camp to Joyce Glacier took about two hours. Ashley and I spent several days up there under unusually gloomy skies scoping out a couple of decent-sized cryolakes at the margins of the glacier before finally puncturing their pristine ice lids with a metal ice drill and inserting instruments to measure light, temperature and oxygen in the water. Within a couple of days, the ice lids had reformed, shielding the melt capsules and anything living within them from the elements. After a month's immersion in the ice-cold waters of the cryolakes, our instruments told a remarkable story. As you'd expect, they confirmed that when it was cold, little sunshine passed through the thick bubbly ice roofs of the lakes – our light sensors showed that as much as 90 per cent of the sun's rays were blocked out.[10] Strangely, though, the dissolved oxygen concentrations in the lakes were close to (and sometimes higher than) what you might find in the air above them – but given that these lakes were cut off from the atmosphere, where was the oxygen coming from?

Through the process of photosynthesis, plants make our planet habitable by taking water from the ground and carbon dioxide from the air, and harnessing solar energy to convert them to organic material, generating oxygen as a by-product. The oxygen in the cryolakes provided a clue that there was life here, in the form of single-celled plant-like organisms – rather like the algae that cause your fish tank to go green if you don't change the water often enough. However, the oxygen in the cryolake waters didn't just go up and up and up, as more and more oxygen was produced by the algae in their sealed abode. It stabilized, which meant that something else had to be taking it out of the water.

Evidently a different group of microorganisms existed in the cryolakes which was consuming the oxygen and regulating its levels in the water. These were the bacteria called hetero-trophs, which can't photosynthesize to create their own food, but instead rely on a supply created by other living things, such as the algae; they use oxygen to break down organic carbon to generate food (and usually energy), and carbon dioxide is released as a by-product. The carbon dioxide can later be used by algae for photosynthesis – so the two types of microbe work together to survive. These clever interactions made the cryo-lake a bit like a terrarium, only substituting the glass with ice and the plants with algae. In this type of enclosed living world, oxygen produced by the plants is used by bacteria, and carbon dioxide produced by bacteria gets used by the plants for photo-synthesis, creating organic material. A system in balance.

One of my research assistants, Liz Bagshaw, a veteran of Dry Valley expeditions, created an artificial cryolake in a freezer back in our Bristol labs by putting some sediment from the bottom of a real cryolake into water-filled glass jars. She was curious to learn whether the very low light levels beneath the cryolakes' thick ice lids affected the balance between the oxygen consumed during the breakdown of organic carbon (by hetero-trophs like bacteria, but also algae) and the oxygen produced when organic carbon was manufactured by photosynthesis (by the algae).[11] If there was, on balance, more 'home-made' living matter created by the algae in cryoconite sediments than could be consumed, then the resulting dark organic material might amplify ice melt on glacier surfaces by darkening the ice, allow-ing it to absorb more of the sun's rays.

Liz ran her experiments for several months. It took a bit of time for the glacial microbes to start to behave like they would in the real world; at first, falling oxygen concentrations sug-gested that the heterotrophs had supremacy over the algae, which were not well established at the sediment surface after their long period of frozen dormancy in a mixed-up mush. After several weeks, though, green-coloured mats of algae appeared on the sediment surface in the jars, and the amount of oxygen dissolved in the water started to rise. The algae now had a strong foothold in their icy world and the heterotrophs deep in the sediments were providing more than enough carbon dioxide to fuel their photosynthetic neighbours in the illuminated surface, generating dark organic matter.

We were fascinated to find that the greater the amount of light, the sooner the moment was reached where the algae were producing more organic material than was being consumed, a result indicated by high oxygen concentrations. But this only happened up to a certain point; if you gave the algae as much light as you might find, say, on the dazzling surface of Joyce Glacier – outside of the cryolakes, that is – their food-making ability was impaired. Interestingly, these algae from the cryolakes of the Dry Valleys were phototrophs that were actually adapted to cope with very little light for photosynthesis and to thrive in the cold. Most living things, including me (ironic, I know), hate the cold and cloudy weather – but in Antarctica and beneath glaciers, it seems, there are organisms that not only love it, but prefer it.

About halfway through our fieldwork on Joyce Glacier, after several weeks of sombre days, the sun finally emerged to blast the surface with its intense rays. I'd expected little to happen, but in fact the whole glacier went wild. Our instruments in the cryolakes showed that the sunshine passing through shot up from 5 per cent to 70 per cent as their ice lids melted and thinned. Suddenly the glacier surface plumbing system stuttered to life – cryolakes grew, streams connecting them were gorged with meltwater and lost their ice roofs, and waterfalls plummeted down over the cliffs and into the main river fed from Joyce Glacier – its flow now a raging torrent which roared towards the ice-covered Lake Colleen.[12]

Such 'glacier flood' events have also been reported else-where in the McMurdo Dry Valleys, which started to exit a decade-long cooling phase around 2002, when there was a large glacier melt event.[13] With this flood of meltwater came a fresh supply of sediments and nutrients – a sort of 'treat' for any life forms downstream from the glacier. What we'd started to real-ize was that these cryoconite holes and cryolakes were food factories. The algae inhabiting the sunlit surface layers of their sediment built up organic material by photosynthesis, bacteria deeper in the sediments 'burnt' it, and this cyclic production and consumption of carbon released nutrients, bound up with the carbon, into the cryolake waters. During cold snaps, the cryolakes and cryoconite holes stored and cycled carbon and nutrients. During big glacier melt events, the larder door was opened and nutrient was transported out beyond the glacier, across the sandy plains.[14] The large lakes downstream from the glaciers in the McMurdo Dry Valleys are depauperate places in terms of the life they can support, since their ice cover can filter out as much as 99 per cent of the light – meaning those extra nutrient supplies from the glacier are very useful for biota in the lakes.[15] Thus, Joyce Glacier was providing life support for ecosystems downstream in Garwood Valley.

Our conception of glaciers was gradually being transformed from cold, sterile deserts to nutrient factories for a whole range of neighbouring ecosystems. This was no 'Valley of the Dead'. Under the glaciers of the Dry Valleys, however, it was a differ-ent story. These glaciers are mostly frozen to their beds, with one known exception, Taylor Glacier, which has ancient brine trapped beneath, the residue of a former ocean, allowing life to be sustained because it's too salty to freeze.[16] For the vast majority of the glaciers here, though, life can only thrive in the surface oases of cryoconite holes and cryolakes – the frozen depths lie dormant.

I always wondered what lay beyond the Dry Valleys, as I gazed up towards the Transantarctic Mountains at the head of Joyce Glacier, transfixed by the epic hunk of ice – the great Antarctic Ice Sheet – that skulked behind them, and pondered exactly what might lurk beneath. The dark, mysterious beds of glaciers and ice sheets had always intrigued me more than their surfaces – an obsession heightened in Antarctica by the utter inaccessibility of a world which had in places been starved of light for thirty million years.

I'm far from the only person to be fascinated by the murky bottom of Antarctica. This is one of the last parts of the world where no land is truly owned, where no wars have happened, and where the environment is fully protected from exploration for oil, gas and minerals. The Antarctic Treaty was signed in 1959 by twelve nations, with the goal of making the continent 'a natural reserve, devoted to peace and science'; today it has more than fifty signatories. Beyond science and tourism, little other activity is permitted. Yet there's a less laudable reason why nations maintain a presence in Antarctica. They are interested in what lies silently beneath the ice – oil, gas, minerals – in case their exploitation becomes practicable in the future. Another draw is the continent's clear skies, which have minimal radio interference, especially within the interior, making them strong potential candidates for long-range surveillance and for positioning systems. Thus science, exploration and geopolitics have been closely interwoven here since the late nineteenth century.[17] Right now, China is in full swing building its fifth scientific research station, with ambitions to build a 'road' to connect them all – essentially a tractor-train iceway from east to west. Over forty nations now operate research stations in Antarctica, which straddle eight territories where national claims were made prior to the Antarctic Treaty (which froze any subsequent claims) by Argentina, Chile, France, New Zealand, Norway, the UK and two by Australia.

The problem with exploring the bed of the Antarctic Ice Sheet is that it's four kilometres deep at its centre, so what lies hidden beneath this enormous frozen blanket is pretty much a mystery. The 1970s saw the advent of intense exploration of the continent from the air by a team of Cambridge geophysicists led by Gordon Robin, later working in collaboration with a Danish team.[18] They flew planes equipped with radar sounders over the ice sheet's surface – radio waves can penetrate ice, and bounce off layers within and below it, creating a phenomenon called a radio echo. By piecing together the echoes from across the ice sheet, scientists could begin to understand its innards. Of course the results showed, as you might expect, layer upon layer of ice, older and older the deeper you went, with sediments trapped beneath the ice. Most revelatory, though, were the large lakes pinned between the ice-sheet base and its rocky substrate, their flat liquid surfaces appearing as bright reflectors in the radar images. One of these was Lake Vostok, a vast body of water deep beneath the ice in East Antarctica – at 250 kilometres long and 50 kilometres wide the sixth largest lake in the world.

The most profound discovery here was that much of the bottom of the Antarctic Ice Sheet was wet – even though, at the surface, the air temperature could be as low as minus 80 degrees Celsius. This is like the giant-skyscraper-ice-cube-in-a-freezer experiment I mentioned a while back, where its immense weight changed the melting point of the ice at the bottom to a little less than zero degrees Celsius. Taking into account a little extra heat from deep in the Earth – geothermal heat – and from the friction of the ice moving over its rocky bed, it makes perfect sense that the bottom of the ice sheet would be watery in places.[19] Since the 1970s, scientists have discovered more than four hundred lakes under the Antarctic Ice Sheet, with rivers flowing between them and swampy zones beyond,[20] uncannily similar to the type of landscape one finds beneath glaciers. The only difference is that in Antarctica, this water doesn't come from the surface of the ice sheet, which is far too cold; it comes from the very slow melting of its sole above its slightly warmer bed. While this may create just a few millimetres of meltwater over a year at any one point, over an entire continent it can amount to a fair quantity – though about one-tenth of the amount produced in Greenland.

I tried to imagine what the bottom of the ice sheet would be like as a home for living things. It would be pitch dark, of course, and there might be a tonne of rock pummelled to pebbles, sand and silt by the glacier as it moved. This would probably be mixed up with a tonne of dead matter pre-dating the ice sheet – trees, shrubs, marine muds, all victims of the ice that amassed on top of them as Antarctica slowly cooled from around fifty million years ago, when carbon dioxide concentrations in Earth's atmosphere fell, until a continent-sized ice sheet formed thirty million or so years ago.[21] Wet, dark, plenty of organic material, scarce oxygen – I thought about other places which had similar ingredients for life – it sounded a bit like a swampy landfill site or, perhaps, the stomach of a cow. That sent my mind spinning, because one form of life that thrives in both these habitats is a certain type of microbe called a methane-maker (technically, a 'methanogen').

Methanogens are a hardy life form. They come into their own when most other microbes, unable to survive with so little oxygen, have quit – indeed, for methanogens, oxygen actually causes them stress. In Svalbard, as you may recall, I'd already found evidence of a certain type of microbe that didn't require oxygen, but could instead use sulphate (which contains four oxygen atoms) to respire organic carbon, creating energy – well, the microbe that steps in once these guys have run out of sulphate is the methanogen. They can thrive in the deepest, darkest places where the oxygen supply has long run out – their resilience has placed them as candidates for life on Mars, which has small amounts of methane in its atmosphere.[22] All that a methanogen needs to flourish is the right type of carbon, some hydrogen and away it goes, producing one of the most potent greenhouse gases on the planet. A release of methane would cause twenty to thirty times more warming than the same amount of carbon dioxide (over 100 years) – which is why cattle farming is so problematic. So, was the bottom of Antarctica like the bowels of a giant cow, building up vast farts of methane in its icy depths? I thought that if I could retrieve some sedimentary mud from beneath the dirty snout of a glacier in the Dry Valleys, it would help me figure out the answer to this gassy question. But somehow I'd need to burrow my way in.

Back home in Bristol, I equipped myself with four shiny new, bright orange chainsaws (two electric and two gas-powered, why stop at one when you can have four?), ear defenders, goggles and hard-wearing safety trousers, then signed up for a chainsaw safety course to make sure I didn't decapitate myself, or anyone else for that matter. One bright spring day I assembled with an instructor, Greg Lis (my comrade-in-arms in Greenland) and Jon Telling (who ran my lab), in the Geography Department car park, where we proceeded to practise slicing through some very large blocks of ice manufactured in one of our walk-in freezers – which was as close as we could get to actually cutting through a glacier.

Subsequently the three of us, with a little help from colleagues, spent over a year chainsawing our way beneath glaciers in Greenland, Svalbard, Norway and, of course, Antarctica. Each of the glaciers rested on very different types of rock. For example, in the Dry Valleys, the glaciers had once upon a time bulldozed their way over gungy lake sediments – full of carbon for any hungry methanogens. In Greenland, the glacier had entombed ancient soils and vegetation in its icy basal layers – remnants of the Arctic tundra, less palatable but still edible for methanogens. Astoundingly, wherever we found a useable form of carbon in our chainsawed ice blocks from the glacier bed, we found methanogens.[23] To find out how much methane they actually produced, we sealed a tiny bit of the dirt from the chainsawed ice blocks in some tiny glass bottles with some meltwater and left them in a fridge.

Two years later we rescued them from the fridge, and learned that the methanogens had indeed produced methane gas, albeit very slowly due to the cold temperatures.[24] Not long after this, an American team drilled through eight hundred metres of ice into one of the subglacial lakes, called Subglacial Lake Whillans, which had been identified beneath the edge of the West Antarctic Ice Sheet. Using a hot-water drill system that could create a borehole larger than a dinner plate, they discovered methane at the bottom of the ice sheet – a lot of it.[25]

The production of so much methane in the icy depths of Antarctica could mean something quite sinister for the planet. Methane is usually a gas, but it's a bit of a shape shifter and can change forms depending on the situation it finds itself in. A certain amount can dissolve in water, but if there's too much methane the water becomes saturated, rather like a sponge that can't hold more water. At this point, gas bubbles usually start to form. But if it's cold and the methane is subjected to a lot of pressure – for instance beneath an ice sheet – it morphs again. Here, a methane molecule takes refuge in a cage of water molecules to form a stable solid that resembles ice and is called methane hydrate (or clathrate). We know that the bottom of the Antarctic Ice Sheet spans valleys and basins, some of them thousands of metres deep, which are packed full of sediment trapped beneath kilometres of ice. These giant vessels of sediment are perfect storage capsules for methane hydrate – deep, cold, remote.[26]

Curious to figure out how much methane hydrate might actually be cached beneath the ice sheet, I started working with two scientists who were equally excited by this quest, Sandra Arndt in Bristol and Slawek Tulaczyk at the University of California, Santa Cruz – they used computer models to replicate major natural systems, such as the flow of ice and how currents move around the oceans. These models are always imperfect, because we have an incomplete understanding of precisely how the natural world works, but they're nonetheless useful for studying vast entities such as the Antarctic Ice Sheet or, indeed, the whole world. It's almost impossible to perform physical experiments on such a scale – but it becomes possible if you can create mathematical equations to describe all the features you think are important about the object of your research, and then tinker with the variables.

As we started to study the methane trapped beneath the Antarctic Ice Sheet, we also began to wonder whether there might be another way in which methane could form beneath the ice – one that didn't need any microbes at all. This was through heat and pressure. Some parts of the ice sheet, particularly in West Antarctica, are geothermal hot spots where the Earth's crust is relatively thin, so more heat is able to seep up from its mantle – giving rise to volcanos, most conspicuously Mount Erebus on Ross Island.[27] Erebus is just one of nearly 140 volcanos identified in Antarctica and which are largely associated with the West Antarctic Rift System[28] – a system of rifts spanning a length of more than 3,000 kilometres between the Ross Sea and the Antarctic Peninsula. Here, the Earth's continental crust has been stretched and linear basins formed as the crust was pulled in different directions. This means that the West Antarctic Rift system, while mostly hidden beneath the ice and similar in size to the more well-known African Rift System, may have the densest cluster of volcanos on Earth. Here, carbon stored in the deep may well heat up and big molecules break into smaller ones, including methane. When we deployed our computer model of the bed of the West Antarctic Ice Sheet to work out how much of this 'hot methane' could be produced along with the microbe-generated methane, we found that it could be billions of tonnes.[29]

The implications of this discovery are terrifying, given that a warming climate may cause parts of the Antarctic Ice Sheet to get thinner or even disappear. The bottom of the West Antarctic Ice Sheet, in particular, sits thousands of metres below the current level of our seas. This means that almost all of its glaciers flow down from the Transantarctic Mountains to float in the ocean. As their ice spreads out across the sea surface, it forms the colossal ice shelves that we see in the Ross Sea, Weddell Sea and many peripheral areas of Antarctica. Pinned in place by rocky mountain slopes at their edges, these floating ice shelves perform an essential role – they press back against the ice sheet, acting like giant brakes that prevent it from flowing uncontrollably into the ocean.

But warm ocean waters are starting to massage the under-belly of these ice shelves in Antarctica, causing them to melt, thin and release more icebergs to the sea.[30] It's not quite as simple as saying that the air warms, therefore the ocean warms, therefore the ice shelves melt. The climate of Antarctica is complex, with teleconnections to different parts of the globe. For example, lying beneath the Antarctic Coastal Current, which moves surface ocean water around Antarctica, are warm, salty deep waters, known as Circumpolar Deep Water, which have been encroaching onto the Antarctic shelf, closer to where the ice shelves hunker. It's not dissimilar to the causes of glacier retreat in Svalbard and parts of Greenland, but the currents driving it are different in each place.

The precise reasons why these warmer waters are appearing around Antarctica are unclear. It may be because deep water is warming up as our oceans absorb more and more heat from the atmosphere, and it may also be because Antarctic winds are shifting.[31] The continent is dominated by the Southern Wester-lies, which swirl clockwise from west to east, driven by the pressure difference between the sub-Antarctic (low pressure) and the zone around 30°S in the sub-tropics (high pressure). Over the past two decades, the Southern Westerlies have become stronger, hugging the Antarctic continent more tightly[32] – this is believed to be due to the growth of the ozone hole and some greenhouse warming, which change the pressure fields.[33] In temperate latitudes, like in Patagonia, the contraction of the westerlies back towards Antarctica is starving glaciers of snow-fall and causing them to shrink. Yet on the icy continent itself, the strengthening of these winds has trapped cold air, meaning that, apart from parts of West Antarctica, air temperatures have not shown such a clear warming pattern as in other glacier-hosting regions of the world, and some locations (mainly East

Antarctica) have even seen periods of cooling in recent decades.[34] (This is another reason why Joyce Glacier in the McMurdo Dry Valleys has been quite stable compared to glaciers elsewhere.) Yet, these subtle shifts in wind patterns around Antarctica seem to be causing warm deep waters to appear closer to the ocean surface, tickling the tongues of ice shelves.[35]

An ice-sheet bed well below sea level, vast ice shelves and balmy oceans – this is not a stable situation. Essentially, as ice shelves melt and thin, the brakes of the Antarctic Ice Sheet are weakening, and ice is flowing faster from the interior to the coast. The glaciers retreat because this flow can't happen fast enough to replace the ice lost at the ice-shelf edge. This is especially true along the west coast of the snake-like arm of the Antarctic Peninsula.[36]

There is one glacier which has felt it particularly hard – Pine Island Glacier, around the Amundsen Sea Embayment. Just over two-thirds the size of the UK, it's the fastest-shrinking glacier on the planet, thinning at more than a metre a year.[37] It has released enormous icebergs some hundreds of kilometres in diameter from its front, since as the ice gets thinner, it gets weaker and cracks more easily and sometimes becomes detached from rocky mountain pedestals on the seafloor that pin it in place. Pine Island Glacier is the biggest loser in terms of mass loss in the whole of Antarctica. Together with its neighbouring ice streams, it holds over a metre of sea-level rise – that's about one-third that of the whole of the West Antarctic Ice Sheet.[38] These Amundsen Sea Embayment glaciers are very important in keeping the West Antarctic Ice Sheet in position, protecting it from collapse. Their future retreat could be key to the future of West Antarctica and our sea levels.

Scientists believe that the collapse of the West Antarctic Ice Sheet might have happened before, during the last interglacial period sandwiched between glacials, called the Eemian, around 120,000 years ago. At the time, global mean air temperatures were around one degree Celsius warmer than today, yet sea levels were at least six metres higher – it's possible that some of this extra water in the oceans was due to loss of ice in Antarctica.[39] Right now, the amount of sea-level rise occurring every year from the melting of Antarctica's ice sheet and glaciers is, at 0.43 mm per year, just a bit more than half that from Greenland (0.77 mm).[40] But if the West Antarctic Ice Sheet were to collapse again it will be much more – say goodbye to East Anglia, low-lying islands like the Maldives, and coastal nations and cities built on reclaimed land, such as the Netherlands, Boston and large swaths of coastal land throughout South East Asia.

As for the methane hydrate – it likes the cold and being under pressure. Remove these conditions and it will become unstable, and when that happens, that solid ice-like methane which has been locked up for maybe millions of years will morph into gas bubbles. We know this happened at the end of the last glacial period in Northern Europe, because scientists have spotted giant craters in the seafloor where beds of methane hydrate became unstable as ice sheets melted away, and erupted explosively as a gas from the bottom of the ocean.[41]

The Paris Agreement of 2015, when nearly every nation pledged to keep global warming to within a total of two degrees Celsius, and ideally one and a half degrees Celsius since pre-industrial times (about the last 150–200 years),[42] failed to take account of any extra warming created by methane emerging from melting ice sheets or permafrost. But if the methane hydrate lies beneath the ice sheets, and it can be released, then we'll possibly need to cut fossil fuel emissions to a much greater degree to keep warming to within the same limits – huge uncertainty surrounds this issue, because no one has produced hard evidence for methane hydrate beneath ice sheets at the present day, only in the past.[43] And even if it is lurking in the Antarctic depths (where we know there's methane in dissolved form), it's possible it might just be converted to the less harmful carbon dioxide before it has a chance to escape into the atmosphere – because another group of microbes beneath the ice sheet obtain their energy by transforming methane into carbon dioxide. (They are called methanotrophs – 'methane nourished'.)[44] How good a job these methanotrophs do here, though, probably depends on how much time they have to get to work on the methane before it is released into the atmosphere.

A PhD student of mine, Guillaume Lamarche-Gagnon, who had arrived from Québec one fine autumn in 2014, beautifully illustrated the importance of the ticking clock when it came to the fate of methane produced beneath ice sheets. He had launched a quest for methane during our field campaign on the Greenland Ice Sheet in 2015. Having grown up in a rural part of Canada with plenty of trees, Guillaume was partial to a spot of chainsaw action. Thus, his first attempt to detect the powerful gas in early spring when Leverett Glacier's river was still frozen solid left him head-first down a deep hole chain-sawed in the river ice, his legs protruding out the top like the flapping tail of a seal. He did indeed detect very high levels of methane in meltwaters at the bottom of the hole, thought to be seeping out from beneath the glacier's snout. Guillaume went on to show that the giant rivers that drained meltwater from the deep, dark underbelly of Leverett Glacier in the peak melt period were also saturated with methane, simply because they ferried it very quickly from the soft sediments at the glacier bed where it was produced to the glacier snout where it was released into the atmosphere – the methane-guzzling methanotrophs didn't have enough time to get to work.[45]

A slightly different situation might arise if methane is released for the very first time at the margins of a retreating ice sheet, for example by what we might call a methane seep – an occurrence where methane escapes from sediments into the ocean above it, normally carried by the upward flow of fluid through rock and sediments (present ice sheets are mostly surrounded by oceans). In 2020 the first methane seep was discovered in Antarctica just off Mount Erebus in the Ross Sea by a group from Oregon State University, who detected high concentrations of methane brought to the seafloor by fluids bleeding up through the sediments.[46] Although unrelated to climate or glaciers, this methane seep told a very interesting story. It was first observed in 2011, but after five years of study, methanotrophs were only just establishing themselves on the sediment surface – taking advantage of all that new methane as an energy source – and they certainly weren't able to oxidize all the methane to carbon dioxide before entering the ocean column. This release of methane to the ocean is not good news; together with Guillaume's findings in Greenland, it seems to suggest that time is of the essence when figuring out how much methane might be released from the bottom of an ice sheet and its muddy surrounds into the ocean or the atmosphere.

When you spend time in places like the McMurdo Dry Valleys, it's incredibly hard to fathom the implications of fast and dramatic change in the polar regions – even as a glaciologist. One is hypnotized into feeling 'out of time', suspended at the apex of a pendulum as it sways rhythmically from one side to the next, as the fleeting summer passes and the winter embraces the land in its cold, dark cloak. Every day, on my trudging journey to the glacier, I would pass the lumpy moraines in front of Joyce Glacier fanning out like ripples on a pond – evidence that the glacier's snout had once advanced forwards, even if it seemed so silent and motionless now. Then there was the testimony of the rocks littered across the proglacial plain, etched by the wind into otherworldly sculptures over millennia. There was one rock which I called 'the laughing man'. It resembled a man's face rising out of the grey sands, turned ninety degrees as if lying on his back, eye sockets hollow, mouth agape as if uttering one final mocking laugh – like a messenger from beyond, reminding me that life was futile and fleeting and that all things change.

Yet a brief period of sunshine after weeks of gloomy days would bring the whole glacier surface plumbing system gurgling to life, sending nutrient-rich water gushing down the Dry Valleys, stimulating life to bloom in ice-covered lakes – and the rate of change would flip from slow to fast. These two time scales are not as irreconcilable as they might sound, for the Earth's natural response to the warming climate encompasses both slow, gradual change (what we often call linear change), and rapid, accelerating (non-linear) shifts characterized by moments of crisis when tipping points are crossed. The great fear is that we're close to a tipping point for the Greenland and Antarctic Ice Sheets.

My time in the McMurdo Dry Valleys certainly served up some harsh lessons about the fragility of life – none more poignant than the story of the hungry little Adélie penguin that stumbled into our camp. Protocol dictated that we could do nothing for him – and eventually, after days of flapping his wings feverishly in a vain attempt to gain our attention, he just gave up. One day, he waddled away. It was partly a relief, as the sight of a poor animal we could not save gnawed at our consciences daily. Then, several days later, while trekking across the lake ice to the glacier, I discovered his tiny corpse just metres from our path. This dear, comical creature, recently so vivacious, was no more, all life extinguished, his sooty eyes dull against his white shroud. Subsequently I watched him diminish every day as the hungry skuas took their feed, limb by limb, my stomach tightening every time we passed, half wanting to avert my gaze, but always looking in the end.

In so many ways, the fate of this single, tiny penguin reminded me of the brevity of life, the sadness that can so quickly consume you without warning. In the end, nature is often the winner – we can slow it down, alter its path for a short while, but ultimately there's little we can do. Despite my agonizing worries, my mother was still with us when I returned from Antarctica, and it would be three years before that particular ending came to pass – another outcome that I was powerless to prevent. Intervention or not, whatever one does or doesn't do, nature in the end simply takes its course.

# PART THREE

# *In the Shadow of Glaciers*

# 5. Beware of the GLOF!

*Patagonia*

In the wilderness, sheltered by just a thin skin of canvas stretched taut a foot or two above your head, you may not sleep a wink, and you may feel spooked by peculiar crashes and rumbles in the darkness – but you'll also feel connected to something much, much bigger than yourself. Who knows what this bigger something is, let alone what it's called? All that matters is that as human beings we seem to crave a connection with it. For me, Patagonia was about reconnecting with myself and with glaciers after a painful few years of disconnection. My first visit was in August 2016, in the depths of the Chilean winter. I'd spent the night in a tiny bivouac tent close to the snout of Steffen Glacier, which protrudes like a long, thin nose out of the bottom of the northernmost of Patagonia's two giant ice fields, dozing fitfully, my mind alert to the percussive beating of the rain.

At dawn, the drumming abated. I peered outside, thinking that the heavens had at last closed their sluice gates, but instead soft flakes of snow were falling silently to the rain-soaked ground. I did not relish getting up. My boots were sodden, and had been for several days, while most of my clothes exhibited varying degrees of dampness. I reluctantly unzipped the door of my tent, to let the outside in and the inside out, and was greeted by a spectacle that made me feel like I'd woken to another world. The snow had failed to settle around our camp, the ground was too warm to enshrine its frozen flakes. A thick blanket of cloud sat heavy upon the tree-clad slopes, but always moving, shape-shifting, its wispy edges playing among the thick forest canopy, like trails of incense in a dimly lit room. I watched, enchanted by this theatre of clouds in motion, their ghostly shapes dancing above an amphitheatre of trees and ice.

Slowly the clouds to the west opened a chink to let through blue light, slowly raising their curtain of rain and snow to reveal great mountains etched in white and black. Across their flanks a clear line was drawn, like a watermark, where snow had turned to rain at low altitudes. Glimpses of smoothly moulded granite bore witness to a much larger ice sheet that had covered Patagonia 20,000 years ago. Yet despite the erosive force of thousands of years of moving ice, the mountain sides were often lumpy, crumpled like used paper bags, capped by gleaming snowdrifts. This apparent contradiction of juxtaposed smooth and roughly hewn rock reflects two important processes by which glaciers sculpt their landscape, which often occur on different sides of a mountain when buried by ice. 'Abrasion', a bit like sanding, occurs on the upstream side as the glacier flows over the obstacle and its base melts under pressure, while 'plucking' (or quarrying) happens on the downstream side, when meltwater worms its way into rock crevices and freezes as pressure is relieved – the rock becomes weakened and fragments are eventually pulled away from the mountain as the ice moves downhill.

After all my travels to icy lands over twenty years, you'd have thought I would be well prepared for Patagonia – but I wasn't. I've never been as cold as I was that first night under canvas on the fringes of the Northern Patagonian Icefield. I'd brought with me a brand new, ludicrously expensive tent which turned out to have zero ventilation capability, its dark sheet stretched over a single hooped pole just above my nostrils as I breathed water vapour in and out through the night; the beads of condensation off its walls thudded arhythmically onto my feather down-filled sleeping bag. Down, it also turned out, was unsuitable for Patagonia – brilliant for Greenland (a classic polar desert) but not for anywhere this humid. Even the new-fangled 'hydrophobic down' in which I'd invested proved hopeless, growing ever heavier during the night as it absorbed the drips from the canvas above. Although I'd packed as many clothes as I could around me for extra warmth, when I finally emerged in the grey dawn I was numb inside and out. A pair of condors circled above, as they often do in winter when there's a chance of living creatures perishing somewhere on the plains – I dare say they had me in their sights.

I fumbled to unearth a small stove and some coffee beans from an aluminium crate, craving a jolt of caffeine to jump-start me from my torpor. One of my companions, Jon Hawkings, sensing my exhaustion, handed me a flimsy sock to use in the absence of a coffee filter, through which the dark liquid dripped until there was sufficient to fill a small cup. I lifted the heavenly smelling brew to my mouth, inhaled deeply, closed my eyes, and savoured the feeling as it brought me back to life again. Thank God, I thought, I may manage to do something useful today after all.

As my cold, damp night testified, Patagonia is a land of water and in all forms. This thin sliver of land, which stretches along the Pacific coast between about 37° and 55° South, strad-dling the borders of Chile and Argentina, spans an area twice the length of New Zealand, which falls within the same latitu-dinal zone and has a similar climate (not to mention glaciers). Patagonia's abundant moisture arrives via the Southern Westerly winds that sweep over the Pacific, delivered to the coast by an endless procession of storms and then, as the moist air rises sharply to the peaks of the Andes, as snow onto two giant ice fields. Together, the Northern and Southern Patagonian Ice Fields form the largest expanse of ice in the Southern Hemisphere outside of Antarctica, draping their icy folds over sharp granite peaks, feeding hundreds of glaciers stretching all the way down to sea level. The Southern Ice Field accounts for more than three-quarters of their total area, and together they hold about forty times more ice than the European Alps.[1] Despite this, due to their remoteness and their challenging weather, we know comparatively little about them.

The amount of precipitation experienced by the coastal region in Patagonia is almost beyond comprehension – from 5,000 to 10,000 millimetres per year (yes, that's a five- to ten-metre layer of water, either from rain or snow).[2] To give this some context, Bristol receives about a tenth of that, and London even less – believe it or not, the UK is a relatively dry nation. The Southern Westerly winds drive the weather in Patagonia and, indeed, the whole Southern Hemisphere. During the height of the last glacial period, 20,000 years ago, a much colder Antarctica pushed these persistent winds further north towards Patagonia, widening the zone over which rain and snow fell and feeding a huge Patagonian Ice Sheet nearly 2,000 kilometres in length. Cooler oceans and air also helped Patagonia's glaciers to grow. At the end of this glacial period, 18,000 years ago, the zone of the Southern Westerly winds shrank back towards Antarctica and triggered the sudden retreat of ice in Patagonia to ultimately form the two smaller ice fields we see today, along with mountain glaciers in the Cordillera Darwin in the far south.[3]

Because of these unrelenting westerlies, Patagonian glaciers receive a gigantic amount of snowfall – a blanket about thirty metres deep – on the western flanks of the ice fields where the moist winds rise as they hit the peaks of the Andes.[4] Like the Haut Glacier d'Arolla in the Swiss Alps, these glaciers are wet at their beds, and are able to slide over a thin layer of lubricating water down to the ocean. This attribute, combined with huge amounts of snowfall, makes them some of the fastest-moving glaciers in the world, virtually sprinting at ten kilometres a year,[5] or some tens of metres a day. They need to flow quickly in order to shift all that snow from top to bottom. It's a bit like squeezing a full tube of toothpaste, the equivalent of adding pressure from ten metres of snow every year – the paste flows rapidly out of the nozzle. Patagonian glacier tongues are ridden with crevasses due to their speedy flow, and their bodies of mashed-up ice extend all the way to the coastal plains, engulfing the dense temperate rainforest. Yet most Patagonian glaciers are not in a good way; over the last fifty years or so, they have lost ice at world-record rates. They are currently losing about one metre of snow and ice over their surfaces per year, which is among the fastest rates of glacier mass loss in the world.[6]

Why, you may ask, are Patagonian glaciers in such trouble? One reason is that they merrily flow all the way down to sea level – so their icy tongues are already sitting in a warm, balmy climate. Those glaciers whose tongues meet the ocean, mostly on the western flank of the Southern Patagonian Icefield, are suffering the most; they don't just lose ice through their surfaces by melting, their tongues also melt in warming ocean waters, icebergs calving off their fronts into the sea, a bit like in Greenland and Antarctica. The Jorge Montt Glacier of the Southern Ice Field (pronounced, gutturally, with the thick accent of southern Patagonia, 'Horhay Mont', not 'George Montay' as I called it when I first arrived in Chile), is the most dramatic example of

this phenomenon. This glacier has retreated more than ten kilometres since the 1980s[7] – ten times the retreat rates seen at the
Haut Glacier d'Arolla or Finsterwalderbreen.

Yet the situation is no rosier on the Argentine side of the
Southern Icefield, and most of the way around the Northern
Icefield, where the glaciers do not reach the ocean and end on
land. The snouts of many of these glaciers now wallow in large
lakes which are forming rapidly in the wake of retreating ice.
Glaciers create perfect conditions for the formation of lakes –
they excavate the terrain as they move, creating hollowed-out
areas. If they stay put for a while, neither advancing nor retreating, all this eroded debris is released from the ice to form a
moraine – a raised ridge of bouldery mud that acts as a dam
behind which meltwater can amass. Patagonian glaciers are
retreating over very flat coastal plains, so meltwater does not
need much encouragement to pond in the hollowed areas and
behind moraine dams. As a result, Patagonia has a higher dens-
ity of lakes than anywhere else in South America, and more
than one thousand new lakes have appeared since the 1980s in
Southern Patagonia, largely due to retreating glaciers.[8]

This big growth in the number of lakes is not just a Patagonian thing – it's also happening in Greenland[9] and in eastern parts
of the Himalaya (Nepal and Bhutan)[10] as the ice wastes back. It's
part of a dismal spiral of glacier decline. Glaciers recede, they
create lakes, their tongues floating in these lakes become
unstable, they release icebergs, the glaciers retreat faster, the
lakes get bigger, and so on and so on. The cycle continues until a
glacier escapes its lake by retreating to higher ground.

The complex interweavings of climate, ice and water created a scientific reason for me to visit Patagonia, but they don't
fully explain how I came to end up shivering in a humble tent
at the edge of Steffen Glacier in the middle of winter in 2016.

I have often since asked myself – did I find Patagonia, or did it find me? Because in truth, it came to me (like many things in life) when I needed it to. I'd been hammering myself between both poles for about a decade, seeking to understand the inner workings of our great ice sheets, pushing to stay at the forefront of the field. As one of very few women leading teams to the edges of our continental ice, I always felt as if I needed to try even harder than my male contemporaries to succeed – so a certain gritting of the teeth was needed.

I'd loved Greenland, and the enormity of the challenges it posed; they'd become something of an addiction for me. Yet, my incessant thirst for more and more extreme experiences in pursuit of answers to seemingly impossible questions meant that the Greenland 'hit' was getting harder to come by. It was super-expensive too – all those helicopters – and had become very busy for scientists. Funding was increasingly challenging to secure as the competition amped up, and rejections to funding proposals came thick and fast. Yet I still had so many unanswered questions from this time, including what would be the impact of the grand landward march of glaciers on life in the oceans and fisheries, and how would it affect humankind? As a scientist, big questions such as these can consume you, sometimes to an extent that is unhealthy.

Then, one day in 2015, the Natural Environment Research Council put out a call for funding bids for joint projects between the UK and Chile, centred upon precisely these issues in one of the world's most fast-changing glacial regions – Patagonia. Brilliant, I thought, before noticing that the deadline was less than three weeks away, and remembering that I knew not a soul in Chile, and next to nothing about Patagonia.

Still, this felt like my last chance. Two years earlier, I had finally lost my mother to cancer, following a seven-year battle, and I'd since been struggling to comprehend both her loss and my new place in the world. At the grand old age of forty-one, for the first time in my life, I was suddenly filled with a deep, uncontrollable urge to have my own family and children – to create a unit around myself. I was in a steady relationship and my glacier work had taken second place for a few years after my mother passed – so it didn't seem a completely incongruous desire. I'd already done the 'get a dog' thing, though, like many things in my life, I did it the hard way, rescuing a half-starved Labrador that had failed on the gun-dog circuit from a farm in Wales. Poppy taught me a great many things, but most of all patience. When I got her, she had serious fear-aggression issues around dogs, men (especially sporting dark jumpers), bicycles and buses, and could not be left in a room on her own – I'd had to take a month off work to save my home from destruction.

Generally, in my life, once I've identified a goal, I'll hurl myself towards it with all the energy I can summon – and thus, shortly before the funding call for work in Chile, I'd found out I was pregnant. Now, though, having always been a free spirit, I found myself filled with terror that I'd never set foot on a glacier again and that I'd lose this central part of my identity. How could I reconcile motherhood with glaciology? I didn't really know, but was suddenly filled with a desperation to secure funding for one last glacier project before having a child on my hands.

I spent a crazed fortnight plaguing friends and colleagues for contacts on the other side of the world, and compulsively reading every scientific paper ever written about Patagonia. The more I learned about this region and its fractured coastline of glacier-tipped fjords, the more I yearned to know how rapid glacier retreat might affect both its spectacular environment and its people. I just about managed to pull together a small band of collaborators in Chile and the UK, and worked on the

13. Antarctica – the loneliest six weeks of my life. Our frugal campsite (including the 'banana' tent) on the shore of Lake Colleen in Garwood Valley.

14. The striking cliffs of Joyce Glacier, with the ice of Lake Colleen in the foreground.

15. Steffen Valley, Patagonia – making friends with the locals.

16. Glamping on the outside, hardcore field camp on the inside – Steffen camp on a rare sunny day.

17. The monstrous tongue of Steffen Glacier wallows in its growing proglacial lake.

18. The start of a new day at Chotta Shigri Base Camp. Note the green beans growing happily in glacial flour.

19. Prof Ramanathan, the two Jons, Monica, Pete and me beside the Chandra River, discussing the nutritious virtues of glacial flour.

20. A place I never wanted to leave – departing Chotta Shigri base camp across the Chandra River.

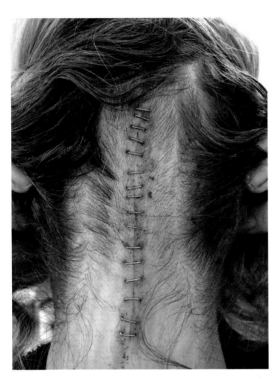

21. A few days after brain surgery, and happy to be alive.

22. The Cordillera Blanca, Peru – back in the wonderful world of water chemistry eight months on from brain surgery.

23. The cathedral-like cliffs of Pastoruri Glacier.

24. So pretty, yet so polluted – Shallap Glacier's lake – pH 3 and rich in heavy metals.

25. Shallap Glacier sliding over its rusty rocks.

26. Becoming the glacier – my role in Erika Stockholm's *The Sad Tale of a Dying Glacier*, performed at the British Embassy in Lima, February 2020.

proposal round the clock. My desperation to win this award further increased after I had a miscarriage at twelve weeks. I entered a tailspin of grief and loss, and my relationship followed my emotions down into the abyss. To escape the hole I was in, I needed to reconnect with the glaciers which had fuelled my passion over twenty years. Fortunately my proposal won funding – in fact it was ranked top of the list – and so off to Patagonia I went.

About the size of Andorra, Steffen Glacier drains the bottom of the Northern Patagonian Ice Field, and was discovered by the German geographer Hans Steffen, who was contracted by the Chilean government in the 1890s to explore areas disputed by Argentina in the Aysén Region; save for a small number of tough Chilean 'settlers' who help manage the forest or earn a crust from passing tourists, the area is uninhabited. The challenge of getting there only added to the glacier's allure for me, requiring (from the UK) two flights to reach Coyhaique in the south of Chile – sixteen hours in the air – followed by two days spent bumping along in a pickup truck on the famous Carretera Austral, which is largely unpaved this far south. Initially constructed in the late 1970s under the dictatorship of Augusto Pinochet to maintain control of the southern territories, the main trunk of this long, twisting road took fifteen years to complete, and reaches the gap between the two icefields; indeed, Pinochet had grand ideas about extending the road across the icefields, but that proved an ambition too far.[11] Today, the Carretera is vital in connecting the two million people scattered across Patagonia, and is one of the most heartstoppingly beautiful highways in the world, weaving its way across more than one thousand kilometres of wilderness. It grazes fjords, channels and gulfs, all lashed by a deluge of rain which brings to life a thick carpet of luxuriant green – from luminous mosses and silvery white old man's beard, to racing green ferns and the soft marl of glacial rivers, set against smudged charcoal skies. ·

I've been fortunate to drive the Carretera Austral most of the way from top to bottom – it's an unforgettable experience. Starting at the busy port city of Puerto Montt in the north, its water's edge stippled with salmon farms, the road skirts the precipitous edge of fjordlands, fringed by thick temperate rain forest which becomes forbiddingly dense as you enter Queulat National Park, where the road winds up and over the mountains. Soon afterwards, you emerge into a zone which has seen much more human influence, with wide stretches of grasslands that are eerily dotted with the remnants of blackened tree stumps dating from the early twentieth century, when the forest was cleared to make way for sheep, cattle and horses. (Cattle farming has since declined in Chile, unlike other South American nations, because its remoteness makes it less economically viable.) South of Coyhaique, you enter the land of ice – the temperature plummets, the road becomes rougher, and towers of granite festooned with snow and glaciers greet you on all sides. Here lies an invisible border between the wet Andes and the terrain east of the mountains, where the air runs out of moisture, the land is parched and rolls out into soft, dreamy grasslands – the Patagonian steppe.

Continuing south, you reach one of the most remote corners of Patagonia, close to its two icefields. Here the roadside is dotted with wooden cabins surrounded by jaunty, criss-crossed fences to create pens for horses and sheep. It was here that we branched off the Carretera to reach the tiny terraced outpost of Tortel, a sleepy village perched on wooden stilts over a fjord; it was only connected to the rest of Chile by the Carretera Austral in 2003. It has no roads, only planked walkways tottering above the water, and is home to some five hundred people who make

their living from fishing and the timber industry, as well as passing tourist trade in summer. It also has a sizeable population of
semi-wild dogs. I met a gorgeous, bouncy collie-cross during
my first visit to Tortel; he was full of fun, knocking me to the
floor in his enthusiasm to play. A year later, I discovered him
limping, with a brutal scar which looked like half of his face had
been devoured. Winter is rough in these parts, the tourist trade
dries up, and food for Tortel's gregarious dogs becomes scarce.
Still, it broke my heart.

I'm not someone who is good at sitting still for long, and the
early stages of this protracted voyage to Steffen, trapped in a
succession of rattling metal shells, drove me slightly mad. But
finally, at Tortel, I was able to step into our small boat on the
Río Baker, gateway to a maze of fjords. Each time I arrived
here, I relished the task of heaving our enormous collection of
bulky metal crates over the side of the boat and onto its wet
boards, nostrils greeted by a fetid aroma of diesel and damp;
my muscles were glad of the exertion after two days spent
folded like human origami in the back of a truck. As soon as
we pulled away from the shore and across the smooth, oily
green waters of the Río Baker, my mood would begin to lift
as my lungs filled with pure, moisture-laden sea air. 'Breathe,
breathe, breathe,' a soft voice would whisper to me, and I'd
perch in the open cabin watching the murky ocean, clouded by
silt from the glaciers, frothing white from the back of the boat
against dark forest-cloaked mountains.

For the final leg of the journey on foot, we were assisted by a
herd of semi-wild horses that carried our gear from the boat to
our camp near the ice. Horses are part of the fabric of Patagonian life, symbolic of this harsh region, and they blend seamlessly
into their windswept landscape, weathering whatever conditions come their way. When not employed, they freely graze

the scrublands in front of the ice, munching on bushes, moss and rough grasses; we would find them wandering among the dwarf birch forest, keenly alert to any disturbance which might signal danger. Their hooves were often split, and they bowed their heads in submission to whatever was requested of them. This saddened me – for some years, I have enjoyed the soul-level connection that occurs between human and horse. I once greeted an animal which was about to carry our gear, a small bay with kind eyes and a softness in his manner; I wondered about his life, where he'd come from, what tales he would tell if only he could speak my language, or indeed I could speak his. I walked over to him and leant my head into his neck; I buried my cold wet nose into his coarse hair, rubbing my fingers gently under his chin, softly humming into his pricked-up ear, which gradually relaxed. He leant into me, I leant into him; I smelled the deep mustiness of his damp coat, felt the gentle raindrops on my face and the warmth of his circulating blood. We were connected in slow motion, with time on pause.

You see, life in Patagonia ambles along slowly. You arrange to do something at ten in the morning, it happens at midday; you plan to drive from one place to another, the road is unpassable due to a landslide. It forces you to let go and roll with whatever prospect appears before you. Unchained from the phone, internet, email; just the present, no future, no past. We get there when we get there. As someone who tends to do too much, too fast, and after the grim year I'd recently suffered, Patagonia was the weight which returned me to equilibrium.

Nonetheless, I had my sights set on a very specific goal, which was to work out how much water was flowing out of Steffen Glacier, and to find out what happened to it all as it poured downriver and into the fjord. Steffen's lake has grown substantially in recent decades, following the retreat of its snout by four kilometres since the 1980s. The glacier is now barely visible from the lake, the view blocked by a large steely expanse of water peppered with icebergs of every shade.

I'll never forget that first mid-winter trip to the margins of Steffen Glacier. Its discomfort was part of its charm; sleep deprivation and cold only heightened my sense of the drama and beauty of this land of water and abysmal weather. Such a prolonged back-country camping mission, I would soon realize, was not 'what one does' in Patagonia on account of the extreme conditions. I'd decided that the best way to investigate the impact of Patagonia's shrinking glaciers and proliferating lakes would be to set up a small camp next to the river draining Steffen Glacier, whose low-slung tongue slinks its way down from the highlands, and which had grown an enormous lake ahead of it during the course of its retreat over just a few decades. I noted my new Chilean collaborators' raised eyebrows while regaling them with our plans – they had deep knowledge of the vile weather we'd probably encounter, of course, whereas we were frankly oblivious.

Luckily a small research institute, the Centro de Investigación en Ecosistemas de la Patagonia, took us under its wing, headed up by its laid-back director, Giovanni Daneri, who helped to arrange the trucks, horses, boats and other essentials for our passage to the camp – and then checked in occasionally to make sure we were still alive. Our team, we later learnt, acquired the affectionate name 'Los Bristols' and folkloric status in this region where people don't usually 'camp' for months on end; we heard reports of locals apparently muttering about us as we moved around with our large, mysterious aluminium boxes and sampling bottles.

The campsite of Los Bristols, where I awoke from my fitful slumber that wet, early morning in August 2016, was a cluster of tents huddled in a sandy dry river channel, fringed by rich temperate rainforest. After my dismal first night under canvas, we swapped our mountain-hardy dome tents of Greenland in subsequent trips for durable Tipis, in order to try to evade the constant rain and so that we could kindle a fire in the winter months. In just a few months, our field camp received about as much rain as the City of Bristol receives in a whole year. From a distance, it resembled a glamping holiday park – until you entered one of the tents, where we'd established a small laboratory, powered by solar panels, or with at least the intention of doing so until we realized that this was a place where the sun seldom made an appearance.

This first time at Steffen Glacier, in the depths of winter, I was astonished to find its river, the Río Huemules (named after a small deer found locally), cheerfully bubbling across the sandy plains and down the steep, smooth bedrock channel by our camp. At every other glacier I've visited, its plumbing system virtually shuts down in winter; its river freezes over, remaining dormant until snowmelt brings it back to life in spring. In Patagonian winter, as I learnt from my first sodden night under canvas, you still have *a lot* of rain, and glacier tongues are so close to sea level that they continue to melt. I hadn't expected this, and it meant that we'd need to measure the volume of the river's flow – its discharge – all year round. Of course, we could not be there all the time; we'd need to rely on instruments placed in the river.

I'd heard alarming tales of Steffen's river being perilous – certainly you would not survive long if you tumbled in. Only a few years earlier, two Chilean scientists had died while attempting to navigate the Río Huemules by small boat for the exact same purpose as us – to install some instruments to measure its flow. My Chilean colleagues were understandably worried by our plans to do the same thing, and when I first set eyes on the river, I did wonder whether we were mad to want to work there.

Yet work there we did, starting with trying to figure out the amount of water in the river. The ideal place to measure its depth was the same spot visited by the ill-fated Chilean team. The sensors needed to be anchored by metal bolts to the channel's sturdy, grey walls of polished granite just over the hill from camp, a job that sounds simple enough. But the temperate rainforest in Patagonia is a force to be reckoned with. One consequence of plentiful precipitation – it gives life in abundance, sustaining a lush, multi-layered canopy of every type of vegetation imaginable, topped by the majestic *Nothofagus*, or Southern Beech. When it rains, the whole forest turns on like a shower, dripping ever-larger droplets of water through its canopy to the sodden earth below.

As a glaciologist, you don't normally have to deal with trees. My first attempt to penetrate the dark mass of the forest was with Jon Hawkings, whose discoveries about iron release from glaciers and its potential for fertilizing the oceans had made him the 'iron-man' of the Greenland Ice Sheet; now he was keen to see whether Patagonia's glaciers were likewise nutrient factories. Luckily, Jon and I found the traces of a path macheted many years ago – it lacked large obstructions like mature trees, but had been suffocated by a jumbled mass of ferns, tangleweed and saplings springing like verdant rockets from the ground. Still, we were quickly disoriented in the mass of green. 'Jemma, I think we've lost the path again,' Jon yelled for the umpteenth time as we hit yet another impasse. This was a mysterious place, where I could imagine stumbling into a hole and emerging somewhere entirely different; I felt like I'd morphed into a burrowing animal, head and shoulders stooped, tunnelling ever forwards in search of another world at the other side of the tree-lined bluff.

I rarely sensed much life in the forest, save for the mass of choking vegetation – but one creature I felt was everywhere. This was a small bird called the chucao tapaculo (*Scelorchilus rubecula*), which is native to Chile and Argentina and likes to skulk in the undergrowth. It resembles a giant European robin, flaunting a bulbous russet breast from which it sends out a cascade of calls to echo around its woody valleys. These haunting sounds – some near, some far – are woven into Patagonia's sensory fabric. The Chilean poet Pablo Neruda beautifully imagined the chucao and its habitat in his collection of poems *Canto General*:[12]

> In the cold multiplied foliage, suddenly
> the chucao's call as if nothing existed
> but that cry of all the wilds combined,
> like that call of all the wet trees.
> The cry, slower and deeper than flight, passed
> trembling and dark over my horse: I stopped –
> where was I? What days were those?
> All that I lived galloping in those
> lost seasons, the world of rain
> on windows, the puma in the elements,
> prowling round with two points of bloody fire,
> and the sea of channels, amid green tunnels
> of drenched beauty, the solitude, my first
> love's kiss beneath the hazel trees,
> all surged up suddenly when in the jungle the cry
> of the chucao passed by with its wet syllables.

Sometimes I was lucky enough to spot a chucao hopping lightly by the path, staring inquisitively at me; but generally, it likes to keep to itself. I came to love this small, bold bird, fur-tively flitting in the undergrowth, but with such a full-throated call you could mistake it for something far grander. As day broke at Steffen camp, I always relished its song, and could almost imagine its words: 'It's morning you know, there may be snow, but you have to go, now to the river . . .'

What lay on the other side of the forested domain of the chucao was almost its antithesis – a bit like a chill-out room after an energetic rave. Smooth bedrock stretched from the forest edge to the riverbanks, like the rounded backs of giant whales basking in the shallows, the water gurgling its way around the horseshoe bend, its slate-grey surface calm in the winter low-flow period. A soft green blanket of moss clung to the upper surfaces of these ice-moulded rocks, disappearing a metre or so above the waterline – bearing witness to the fact that when the glacier is melting faster, the river flows much higher and moss spores are swept away, unable to colonize these sand-blasted shores. This is why we needed to install our instruments in winter – because in summer everything would be underwater.

Compared to wrestling our way through the forest, our work here was a breeze, involving hours spent peacefully wad-ing around, clutching a spanner and a rock drill to attach the sensors to the whalebacks just below the waterline. These scientific instruments were self-contained devices that took readings every half hour or so, resembling elongated metal bullets, about a foot long; they housed a battery, logging device and different types of sensors, some to measure the river's temperature and discharge, and others to record the sediment and dissolved chemicals in the water. The following year, when we returned to retrieve the data from the sensors, we found that the flow in winter time – just as the mosses had hinted – was about a quarter that in summer. Even at its lowest, it was still pretty high for a glacier river – about the same volume as the River Thames in London, bumbling along at a fairly constant level, nourished by the driving rain and drip feed of glacier melt. In summer, however, things sometimes went wild.

I'd heard rumours of something called a 'GLOF' – the name always amused me, conjuring up an image of some kind of rock-dwelling goblin with a gnarly face and a volatile temper. Beware of the GLOF! As it happens, a GLOF is indeed something to be wary of – the initials stand for Glacier Lake Outburst Flood, a phenomenon which has become all too common in retreating glacier landscapes where meltwater is ponding. These lakes are inherently unstable; often at least one of their banks is ice, and the others the rocky rubble of moraines. When full, lakes often breach their icy banks and their waters drain catastrophically, wreaking havoc on rural communities.

People had mentioned that there were lakes in the Steffen Glacier area which GLOFed (to coin a phrase), causing river levels to rise many metres, sweeping up everything in their path; there were alarming reports of local settlers finding their sheep stuck in trees once the floodwaters had passed. It sounded terrifying. We later discovered that our campsite, which rested in a seemingly perfect spot, on flat, sandy terrain, was actually a GLOF overflow channel – but luckily it had become redundant as the main channel eroded downwards. Despite hunkering in tents on the banks of the Río Huemules for many months in both winter and summer, we never experienced a GLOF – but our instruments in the river did. They showed that a couple of these mad floods could happen in a summer, and overnight the river went from the size of the Thames to something more than fifty times bigger. The scale of these GLOFs was almost unbelievable – where were they coming from?

We'd set up cameras high on rocks overlooking the river, well out of the reach of a GLOF, in order to capture images of these elusive inundations. The photos showed, one day, the river placidly delivering its watery load downstream; the next day, it had advanced up its banks by eight metres or so, turbulent eddies swashing violently around its bends, a crazy writhing mass of water. By cross-referring to satellite pictures taken at the same time, we could locate the source of all this extra water. Ten kilometres up the steep valley, the meltwater in a lake pinned at the edge of Steffen Glacier had built up until it was full, forcing the ice to give way; the lake was approaching the size of Derwent Water in the English Lake District, and you can imagine that if that body of water suddenly decided to empty itself out, the Cumbria Tourist Board might have something to say about it. Another time, a GLOF had come down from a different lake, on the eastern side of the glacier. It had flooded its entire river valley, which was transformed into a glassy sheen of water dotted with islands of trees, inundating the wooden houses of the settlers.

Patagonia is not heavily populated, so the scale of any devastation caused by a GLOF is generally small. But in more highly populated parts of the world, these freak floods pose a terrifying threat to local people. One of the biggest GLOF disasters ever recorded was in 1941, in the Cordillera Blanca of Peru. In this particular tragedy a lake called Palcacocha, nestled behind the moraines of its glacier, became so full that its fifty-metre-high moraine dam failed, unleashing an enormous flood wave that destroyed the local town of Huaraz, killing several thousand people. Since then, Peru has invested heavily in structural engineering solutions, including dam walls to keep the waters in and drainage pipes to siphon off excess meltwater, as a result of which there have been fewer GLOFs and no fatalities.[13]

On a global scale, it's not quite clear whether the problem of the GLOFs is getting better or worse. Many GLOFs were observed in the 1930s, as glaciers crept back off their terminal moraines following (natural) warming at the end of the Little Ice Age in the late nineteenth century, and lakes started to grow.[14] Since then, GLOF occurrence has fallen, but this may partly reflect human intervention and the implementation of risk-reduction measures, as in Peru. It's certainly true that lakes are growing in valleys where glaciers are retreating; so finding ways to forecast when GLOFs are going to happen, and how bad they'll be, is critical. Since recording the GLOFs at Steffen Glacier, our small instruments positioned below the waterline in the area of the whalebacks are now being used by the local water-management authorities to help learn about (and perhaps predict) these extreme events; even so, while they're certainly dramatic, GLOFs only account for about one-tenth of the annual flow of the Río Huemules, most of which is just regular rainfall or icemelt.

One of my favourite spots was on the lip of the large moraine in front of Steffen Glacier, framing its lake like a wide grin around its southern shoreline. This accumulation of debris marks the maximum position of the glacier during the Little Ice Age, which occurred in Patagonia around 1870.[15] Clambering up the moraine, the glacier announced its presence with chilling katabatic winds howling off the ice; from here, I could see its fast-moving tongue, its medley of crevasses giving birth to schools of gleaming icebergs, stranded in the still waters, and ponder the ice rivers of Patagonia.

Lakes now fringe much of the Northern Patagonian Icefield, their milky waters lapping frozen glacier tongues and offering a cushioned resting place for icebergs. In many ways this icefield, which is now stranded on land, provides a vision of the future for its neighbouring Southern Icefield as well as glaciers in Greenland, Alaska and Antarctica, whose toes currently dip in the ocean but which are retreating inland as the ocean warms. My studies in Greenland and Antarctica had taught me that the rivers and icebergs that emerged from glaciers were packed with organic carbon and nutrients like phosphorus, silicon and iron, which were either dissolved in the flowing water or bundled up with very fine floating rock particles eroded from the glacier bed. They'd also shown me that these tiny grains of silt and sand travelled from the land, through the fjords, and even out into the wide open ocean, where they could nourish phytoplankton, the tiny plants of the ocean. Was this also true in Patagonia, and what difference would the vast and expanding network of lakes here make to this conveyor belt of nutritious produce?

The answer to these questions takes us back to Svalbard and the idea of the chemical memory of water. You see, if meltwater picks up traces of chemicals and sediment on its passage beneath a glacier, creating a 'chemical memory' for itself, then these lakes help this glacial meltwater to 'forget' where it's been. While trapped by the lakes – a stay that can vary from days in small lakes to weeks or months for a lake as large as the one at Steffen – particles ground up from the glacier bed are lost from the waters, as they settle down in layer upon layer of very fine mud. Such lakes act as a giant water-filter, so that the river loses its sediment and chemicals as its waters travel through them, emerging transformed at the other end. The river waters flowing out of the lake at Steffen Glacier contained one-tenth the amount of sediment particles in a litre of water than rivers flowing from other glaciers I'd studied.[16]

You might initially think this was a good thing for Patagonia's fjords – as we know from Greenland, particles in surface waters block the supply of light to the little food-makers, the phytoplankton. Unfortunately, the lakes do not quite do a good enough job of reducing the shading effect of particles, and the river waters are still cloudy. This is partly because the tiniest particles take longer to sink and are not so fast 'forgotten', so they can still cause a fair bit of shading in fjord waters. Additionally, the ever-rising volume of milky meltwater flow from Patagonian glaciers has been outpacing the trapping of particles in lakes.[17] You might imagine that the growth of the giant lake in front of Steffen Glacier since the 1980s would boost the effect of the lake particle trap, but on the contrary – over this time period, the delivery of fine particles by the Río Huemules to the fjord downstream has risen as more and more water comes off the glacier. As in Greenland, more melt equals more particles, with or without lakes.

Moving from the deglaciated north of Patagonia to the ice-rich south, the growth of phytoplankton in the surface waters of its fjords becomes ever more depressed, as glacier melt feeds its fine particles into rivers, creating a layer of cloudy freshwater which lolls on top of salty ocean water.[18] This freshwater is also poor in the nitrogen needed by phytoplankton; it is the ocean waters trapped beneath it which contain the nitrogen. Thus nutrients and light can be scarce for phytoplankton in the fjords of southern Patagonia, in turn affecting the type and abundance of life. Smaller types of phytoplankton prosper here, as they are better able to survive with low levels of nutrition.[19] In these southerly climes, one might imagine that there's a lot less food available to support bigger creatures further up the food chain like fish. Yet as you nose your way out of Patagonia's fjords and into the wide open ocean, and the particles drop away, scientists have observed larger phytoplankton appearing when ocean waters are examined under a microscope, sustained by more light and by a mix of nutrients from the ocean and the rivers. One nutrient in particular is generously supplied by Patagonia's rivers – silicon.[20] This is an important food for a large type of ocean-dwelling phytoplankton called the diatom, which uses the silicon to build beautifully intricate glass shells around its single-celled body.

So Patagonia in some ways resembles Greenland – melting glaciers shape the entire base of the food chain of its fjords. But what will happen as our climate continues to warm? On the one hand, glacier melt is increasing; in areas where this is a major water source for rivers, streamflow has risen in recent decades as air temperatures have warmed.[21] On the other hand, Patagonian rivers which are less reliant on glacier melt are drying out as the tracks of moisture-packed Southern Westerlies shift poleward, probably as a result of human-induced changes to our atmosphere.[22] The downstream sections of Patagonia's largest river, the Río Baker, whose murky green waters pass by Tortel, has seen its flows fall by about one-fifth since the 1980s.[23] Overall it's a confusing story, which makes it challenging to work out what will happen next. But in a region where the vigorous cycle of water is critical to all forms of life, one thing is certain – everything is connected up.

As with all my glacier expeditions, I'd gone to Patagonia seeking answers; and while I found some, these generated even more questions. Even in such a remote, hostile region, seemingly at the end of the Earth, the effects of our warming climate were plain to see. Glaciers were turning sickly, their icy tongues wasting away at record speed, and the fjords and all the life they harboured were going to pay for it. In fact, we were all going to pay for it.

On one of my last visits to this rain-washed land, as if mimicking the trials of its suffering glaciers, I started to sense changes in the health of my own body. It was a short trip at the end of October 2018 to download some data off our instruments in the river. Unusually for Patagonia, the skies were clear, the sun beat down on our sandy campsite, and I was overjoyed to be back in this place which had so generously brought me reconnection. I was better prepared than my first time there – I'd brought a tent which breathed and a sleeping bag that did not absorb moisture. Even so, everything felt a struggle, and I couldn't work out why.

For a year or so, I'd been having painful headaches, caused (I was pretty sure) by my incorrigible habit of using my laptop in the most ergonomically disastrous places – planes, trains, the sofa, the sofa again. I was a workaholic, and most workaholics work wherever and whenever they can. But now, at Steffen Glacier, the pain was worse. I had to take down my tent on the last day, but the act of crouching on the ground left me in agony as searing pain shot up into my head. I ended up crawling in the dirt like a creature about to expire, pulling out one tent peg, then lying flat while my head pounded and I gathered energy for the next one. This simple task took an hour, and by the end I could barely stand. Ah, that laptop, I berated myself – you need to sort it out, Jemma.

For the first time in my life, the thought of walking out with a full rucksack seemed a step beyond manageable. So, I was forced to do the one thing that my Chilean colleagues had warned me never, ever to do. I caught a boat down the Río Huemules. It was a calculated risk – this was a river that was freezing cold and dangerous, with rapids and sections of white water, and it had already claimed lives. I sent my two companions, Jon Hawkings and Alejandra Urra, one of my PhD students, on foot to the pickup point on the coast, two hours' walk away, and awaited the small boat. It arrived, humming around the river bend, a black inflatable rubber vessel full of people, including the local ranger Don Efraín, whose chiselled, weather-blasted face I carry in my mind as the countenance of Patagonia. I breathed a sigh of relief as I settled down into the bows, content simply to watch as we glided through the chalky waters, past wooded banks I'd never seen before, down frothy rapids, only pausing briefly to machete some stalks of the nalca (*Gunnera tinctoria*), an edible plant resembling a giant spiky rhubarb. I had no idea what to do with it when offered some, but learnt that the trick was to use your teeth to strip off the outer gnarly layers, then chomp on the bitter-tasting flesh inside, while juices dribbled down your chin.

The freedom of just letting go, allowing the waters to carry me and trusting in the skill of the locals, was perhaps a valuable lesson for life. I did make it safely out of Patagonia, though on that occasion we arrived by boat in Tortel in gale-force winds to find our pickup truck unuseable due to a puncture – another problem to fix while we stood exhausted in the hammering rain. The long flight home came as a respite, now that the pain in my head had abated somewhat. But it turned out that I'd breathed my sigh of relief too soon. On landing at London Heathrow, I got up from my seat, only to black out in agony before I could even exit the aircraft. I still had no idea why – maybe I was a bit stressed? I decided not to worry about it. The true reason would not emerge until a month later.

# 6. White Rivers Running Dry

## The Indian Himalaya

The wind whisked up flurries of snow, which danced like a thousand winged insects against the dark, foreboding mountain flanks while the hypnotic chants of our Nepali guides rang out across the valley; they'd burst into song while pausing to rest against a giant sand-coloured boulder. I admired them – wearing no more than a pair of trainers, chinos and a thin outer jacket, they hopped up the glacier as if ascending a short flight of stairs. Meanwhile I stood motionless, running my eyes over the serrated pinnacles towering above us – at some 6,000 metres high, almost twice the height of anything I'd seen in the Alps – and shrinking into my thick woollen scarf in a futile attempt to keep the biting cold at bay.

I was high in the Western Indian Himalaya, named after the Sanskrit for 'abode' (*alayah*) of the 'snow' (*hima*), part of three intersecting ranges, the Hindu Kush, Karakoram and Himalaya. This vast mountain range splashes a wide arc of meringue-topped peaks across Asia, starting with Afghanistan and Tajikistan, through Pakistan, India, Nepal, Myanmar, Bangladesh and Bhutan, before terminating in the high Tibetan Plateau of China. In the west, the Karakoram mountains are blasted by the westerly winds that dominate weather patterns throughout the mid-latitudes of the northern hemisphere, delivering rain and snow mostly in winter. Yet travelling eastwards, you enter the realm of the powerful monsoon which drenches India, Nepal and other

Asian countries every summer. The Himalayan mountain belt contributes by blocking out cooler polar air to the north, which causes the Indian continent to heat up furiously in summer. It's the heat difference between the hot land and the cool Indian Ocean that draws warm, wet sea winds into India, until these heavy clouds burst, often while rising over mountains. By way of contrast, the regions to the north, such as the Mongolian Steppe and the desert province of Xinjiang in western China, are starved of precipitation, making them dry beyond belief.

The Himalaya and its aligned mountain ranges (referred to subsequently as 'the Himalaya' for brevity), which is the greatest glacier-covered region on the planet outside of the poles, is often nicknamed the 'Third Pole'. Unlike Greenland and Antarctica, where huge ice sheets smother the underlying peaks and troughs, this mountain range is host to over 50,000 valley glaciers, creeping their way down from summits which pierce the skies at up to 8,849 metres – the height of Everest. These 'Water Towers' of Asia are a vital resource in the region, storing this most fundamental of substances in frozen, solid form. In springtime the blanket of snow that wraps the mountains melts slowly, and in summer, once the snows have gone, glacier melt provides a steady drip feed of water into rivers. In the central and eastern Himalaya, bursts of summer monsoon rains top up the water supply on lower mountain slopes, turning to snow higher up on the glaciers, which helps them stay healthy.[1] Meltwater and rainfall also seep into the ground, which soaks them up like a sponge, later flowing out of the hillsides via springs.[2] Thus, even during a protracted period without rain, melt from the trusty glaciers will continue to provide water downstream.

So the Himalayan mountains work rather like a series of 'taps' which can turn on and off at different times, the water supplied by them eventually coalescing to feed ten vast rivers that traverse entire nations. Most notable in India are the Indus, Ganges and Brahmaputra, which nourish the world's most extensive irrigated area of agriculture and permit people to survive frugally in the most rugged and remote of valleys.[3] In the Himalayan region, the story of water underpins almost everything – agriculture, politics, even spirituality and religion – and glaciers sit very much at its heart. However, the air around the glaciers in the Himalaya has warmed by about 0.2 degrees Celsius per decade over the last fifty years – which may not seem much, but should this continue throughout the century, it will carry us beyond the pledge of the Paris Agreement of 2015 as to the upper limit on total climate warming. What will this mean for the nearly quarter of a billion people in the mountains, rising to over a billion people if you include those living on the plains, who are in some way dependent on the meltwater?[4]

This was my first time in the Himalaya, and the result of a stroke of what you might call luck, but which I rather think of as magic. One day in autumn 2016 an email popped into my inbox from Professor Al Ramanathan of Jawaharlal Nehru University in Delhi – someone I'd never met before, though I'd once hosted one of his PhD students at my lab in the UK. He'd spotted an announcement for funding by the Indian government, uniting with UK funding agencies, and was wondering whether I might like to join forces? It was a little like when I'd applied for funding in Patagonia, but this time instead of three weeks, the deadline for proposals was just *one* week away. How could I possibly rustle up a twenty-page proposal in so little time, I asked myself? But how could I pass up such an opportunity? I had to go for it – so I shut myself away and plunged into a manic twenty-four hours of research and writing. I sent out a flurry of urgent emails to trusty glaciological comrades to see if they might like to join this impromptu venture – almost instantaneous replies arrived back in my inbox from my old partner-in-crime Pete Nienow, alongside chainsaw accomplice Jon Telling (now a lecturer at Newcastle University), 'iron-man' Jon Hawkings and sensor developer extraordinaire Matt Mowlem. We dashed off the proposal in the blink of an eye – six months later, as if by some miracle, it was awarded funding.

Almost a year later, in September 2017, we landed in Delhi. The group comprised myself, both Jons and Pete, Alex Beaton (one of Matt Mowlem's researchers who had worked with us in Greenland), Andrew Tedstone (a former PhD student of Pete's) and Sarah Tingey (Bristol PhD student and all-round mountain enthusiast). It felt like a reunion of friends, drawn together from disparate places and various eras of glacier study; but this was very different from our prior adventures. We were immediately blasted by the stifling heat of Delhi, the confusing blend of smells, from scorched tarmac and putrid drains to spiced food and petrol fumes, the streets thronged with tooting cars and rickshaws – being used to freezing places totally uninhabited by people, this was sensory overload for me.

The next morning, Professor Ramanathan stopped by our urban-chic guesthouse that nestled into the walls of a narrow side street in the trendy district of Hauz Khas. He appeared in the lobby grinning from ear to ear – after many months of frantic emails back and forth, his warmth and confidence felt immediately reassuring. Prof. Rama (as he came to be called – the persistent use of titles is common in Indian science) was accompanied by his right-hand man and logistics supremo, Mr Chatterjee. The Indian team contributed decades of knowledge from working on remote Himalayan glaciers, together with a knowledge of how these glaciers behaved in terms of their plumbing, health and flow, to the collaborative venture; the British team pitched in with new skills around the understanding of how glaciers might host microbial life, and release carbon and nutrients to lakes and rivers downstream. Uniting these different strengths was almost unprecedented at the time in this part of the Indian Himalaya, and our brief trip there was a test drive to see whether we could make something work in the longer term.

The first error we discovered, at our initial meeting with Professor Ramanathan and Mr Chatterjee, was that we'd brought far too much equipment – the small plane that would take us from Delhi to the small mountain town of Manali in Himachal Pradesh only allowed one standard twenty-kilo bag per person. It was a very different story to the tonne of gear we were able to pile into the back of a helicopter in Greenland. The solution: Jon Telling and Sarah secured a car with a driver, packed our surplus stuff in the back, and set off just after midnight on the fourteen-hour journey by road to Manali, thus salvaging our field campaign. Once in Manali, we met the rest of the Indian team; most notably new-start PhD student Monica Sharma, who had never been in the high mountains, let alone on a glacier, and Masters student Som Mishra, who already had several glacier expeditions under his belt. We got off to a faltering start with the wider team, because the place I had booked for us all to stay was (in Mr Chatterjee's words) 'an Iranian drug den' and 'terrible place'. This was a quirky hostel in the quiet, old part of town, with tiny cobbled streets lined by unruly growths of marijuana plants. The smell alone was intoxicating. A sign on the front wall of the hostel announced, in flamboyant lettering: *We are all mad here*. Maybe for the first time ever, I felt I'd come to the right place. But it didn't suit our Indian colleagues, who decided to stay elsewhere . . . first diplomatic mission failed.

From Manali all ten of our Indo-UK team piled into a truck – rucksacks and crates strapped precariously to the roof – and joined the painfully slow procession of military vehicles and trucks nosing their way up to the Rohtang Pass, close to 4,000 metres above sea level, and only open for a few months in summer. The rest of the year, the weather can be vicious – perhaps accounting for its name, which means 'pile of dead bodies' in Persian/Farsi, a grim reminder of all those who have frozen to death there. The Rohtang Pass is a cultural watershed; to the south lies the Kullu Valley, where Hinduism prevails, while to the north are the Buddhist valleys of Spiti and Lahaul, where we were headed.

Brightly coloured Tibetan prayer flags at the pass signalled this shift – not just a religious one, but also in the thinness of the air, which I started to feel for the first time as I walked across the gravel to try to take a photo that wasn't clogged with chattering tourists. Just a few strides from the truck, I was already out of breath, while the cold, dry air caught the back of my throat, making me cough. Beyond Rohtang, the road was treacherous – a succession of hairpin bends, mostly bounded by a sheer precipice on one side or the other, it was riddled with holes and bumps which left us shaken in more ways than one. It was here, about a year before my blackout on the plane following my 2018 visit to Steffen Glacier, that my head started to ache as it jerked to left and right in time to the jolts of the truck. By the end of our rattling drive, it felt like it was about to explode. 'Must be the altitude,' I told myself.

Our arrival at the final stopping point along the road for Chhotta Shigri, the glacier where we were headed, was a moment of great relief. If I stood upright, then perhaps the pain in my head and neck would go away. Even so, I felt some apprehension – for I'd heard unnerving rumours that the passage to the glacier base camp required one to cross a steep-sided river gorge . . . suspended in a small metal box. Our Indian friends laughed as they watched us size up the crate, its steel arms connected to a metal wire which stretched from one side of the chasm to the other. Figuring that I might as well get the ordeal over with, I hopped in, trying (but probably failing) to look confident.

Moments later, I was swinging like a solitary animal in its cage over the deep gorge below. As the iron box glided spasmodically from one river bank to the other, I inhaled long and hard and surrendered control. 'It's all in the mind,' I yelled to the others when I reached the other side – but I wasn't entirely convinced by my own bluster. I'm not someone who is afraid of heights, having spent a good fifteen years rock climbing. But while climbing is about taking control – deciding on your line and your gear, when to rest and when to venture upwards – this river crossing was about ceding control, which is an entirely different sensation.

After a short, painfully slow uphill plod from the river crossing, we arrived at the Chhotta Shigri base camp, the tiny permanent research station established by Professor Ramanathan to study the glacier at the head of the valley. Chhotta Shigri means 'Little Glacier' in the local Lahaul dialect. It's not that small, though – around nine kilometres long, about twice the size of the Haut Glacier d'Arolla in Switzerland – and sits adjacent to one of the Himalaya's longest glaciers, Bara Shigri ('Big Glacier'). Together they discharge their meltwaters into the Chandra River, across which we'd swung in the metal cage. Butting up against the borderlands of Kashmir to the north and Tibet to the east, both glaciers are neighbours to territory disputed with Pakistan and China respectively, and as a scientist you need permission from India's ministries of both external and internal affairs before going there. (In practice, you hear if you have permission, you don't hear if it is declined, and there's always a long, painful silence either way, making it difficult to plan anything.) Sarah and I were apparently the first Western women ever to have been allowed in.

It was the tale of Chhotta Shigri's water which had first piqued my interest – for its voyage is long and tortuous. The Chandra River, ferrying away melt from the glacier, later joins the Chenab River, which crosses into Jammu and Kashmir and ultimately finds its fate on the plains of the Punjab in Pakistan, emptying into the mighty River Indus. The Indus is significant, since it feeds parts of Pakistan where the demand for water far exceeds its availability – for this is the most water-stressed region in the world.[5] The upper part of the Indus basin supplies one of the greatest irrigation systems in existence, linked to several large dams built by Pakistan. Somewhat confusingly, the waters of the Chenab are sourced in the Indian Himalaya, but they are regulated by Pakistan, according to the Indus Waters Treaty drafted in 1960[6] – an early example of international water sharing, whereby India gained control over three eastern rivers and Pakistan over three western rivers, including the Indus and Chenab. It's not a perfect deal – as you might imagine, it's a little hard to put boundaries along something that flows.

The political struggle to control the region's water has underpinned countless conflicts in one way or another. It's one of the reasons for the dispute between India and Pakistan in the Indian-administered state of Kashmir, since rivers here are the origin of Pakistan's precious water. Pakistan already maintains several dams for hydro-electric power on the Chenab River. India has plans to build more, which is permitted under the treaty, so long as it doesn't affect the water supply downstream in Pakistan. Pitched as a solution to a rising population and water scarcity, this strategy has political undertones. If India develops hydro-electric power projects on the Chenab, it will suddenly be in a position to control water supplies to the heavily disputed Kashmir, and thus to Pakistan.[7]

Thanks to Professor Ramanathan and his team, Chhotta Shigri is the only glacier in the Upper Indus Basin with the logistical infrastructure that has allowed scientists to study its flow of water over two decades. Along with its icy neighbours, this glacier is pivotal to the struggles around water supply in the region; of the many great Asian rivers which find their source in the high Himalaya, the Indus is the most heavily charged by glacier meltwater. A weighty 40 per cent of river water in the upper Indus basin, and up to 90 per cent in some mountain tributaries, comprises meltwater which arrives in spring and summer when rainwater becomes scarce in this westerly region.[8] The much more monsoonal upper Ganges basin, on the other hand, only has about 10 per cent of its water as meltwater, simply because there is more rainfall.[9] In all rivers sourced in the Himalaya, the higher you go up into the mountains, the more weighty glacial meltwater becomes as a proportion of the river's water supply. It's notable that the people who are most vulnerable, whose farms and villages hug remote, steep-sided valleys in the high Himalaya, also rely the most on glacier melt for their water.

Himalayan glaciers are a varied bunch, coming in all shapes and sizes, and behaving differently from each other. Some of them trap snow on their frozen bodies in winter, some in summer via the monsoon, and some (like Chhotta Shigri) in both seasons. Some end on land, while others have tongues that wallow in lakes. Unlike most glaciers in the Alps, the surfaces of Himalayan glaciers are also extremely dirty, often mantled by a layer of coarse rock that has fallen there from on high.

Add to this the fact that these glaciers are stretched over a mountain belt spanning more than 3,000 kilometres, where fieldwork involves dealing with some serious weather and logistics, and it becomes obvious why keeping track of Himalayan glaciers is challenging.

Chhota Shigri's snowfall mostly arrives on the westerly winds, since the glacier sits in the rain shadow of moisture-laden winds of the Asian Monsoon which come from the south. Despite this, however, the monsoon still plays its part in the health of this little glacier since it provides intermittent snowfall in summer that dampens glacier melting by creating a reflective blanket.[10] Chhota is a temperate glacier with abundant water flowing at its bed, and it mostly melts in summer, a bit like the Haut Glacier d'Arolla in the Alps. This was the first small mountain glacier I'd worked on since my time at Arolla twenty years earlier – in many ways, I felt right at home.

Still, my first foray up Chhota Shigri – a slow plod enlivened by the sonorous chants of our Nepali guides – brought home to me the enormity of the task. It took most of the day to ascend from our base camp in front of the glacier, at around 4,000 metres above sea level, up to the mid-section of the glacier 1,000 metres higher. The glacier's rocky forefields were but a benign foretaste of what lay ahead, in particular the gigantic block field of rocks, in some places the size of cars, which armoured the glacier front for several kilometres upwards of its barely recognizable, dirty brown snout.

The reason for all this debris is that the high Himalaya experiences some of the fastest rates of erosion in the world, which ramped up fifty million years ago when the Indian tectonic plate collided with the Eurasian plate, crunching up everything between them to create dizzying peaks which continue to rise by a centimetre a year. This upward movement is tempered by the wind, rain and snow, which strip away layers of rock, raining down detritus into glaciated valleys, some of which will be subsumed by the moving ice until it eventually emerges in the lower part of the glacier's tongue to accumulate into a thick rocky layer – a bit like a moraine, but forming a sea of rocky debris rather than a distinct, elongated cluster (in other words, a moraine). These rocky blankets cover up to a quarter of Himalayan glacier surface area and seem to be thickening as glaciers melt and retreat.[11]

Negotiating this terrain was utterly exhausting, as we picked our way through a chaos of rock, ice and water. This was my first experience of working at such altitude – a staggering 5,000 metres above sea level – and walking required intense concentration. Every few breaths I would open my mouth as wide as possible, almost as if yawning, trying to inhale the absolute maximum amount of air to transmit oxygen to my weary limbs. The skies were brooding to the south, and at some point snowflakes began swirling down. I was starting to suffer, not so much now in my legs but in my head, which was pounding with pain. For a while I had to lie horizontal on the rough ice in order to ease the throbbing. *Why is this such agony?*, I kept asking myself. *No one else seems to be suffering.* My thinking was thick and fuzzy, and decision-making became a real problem as I tried to devise a strategy for sampling snow and ice. My struggle was so obvious that Jon Telling shouted out across the ice: 'Jemma, it's a good thing you're not like this back home, you'd never have made professor!'

After enduring our long bumpy journey, treacherous river crossing and several nights under the stars, relations between the Indian and British team members soon warmed up. Som and Monica became our cheery companions as we assailed the steep glacier front while gasping for oxygen; and at the end of the day, as we sat cross-legged in a circle in the mess tent scooping rice and dal into our hungry mouths, they were surprised to discover that our palates were tougher than they anticipated, too, their eyes widening as we ladled hot spicy chilli pickle onto our dal, which had been toned down to suit bland Western tastes. Jon Hawkings was renamed 'Long John' on account of his tall, skinny frame. Jon Telling became fondly known as 'Bean Man', after recounting the story of how, during our bitterly cold 2015 Greenland field season, he'd encouraged everyone to take a can of tinned food to bed with them so they had something that wasn't frozen to eat at breakfast – his bedfellow was a tin of rather disgusting broad beans. The escapades of Long John and Bean Man, often seen together heading out to sample melt-waters, were the source of frequent mirth within the camp.

Many studies of the growth and shrinkage of individual Himalayan glaciers have taken place in recent decades – but such is the variety of these icy entities that reports have often reached quite different conclusions. The past and future health of glaciers has been considered as part of the Intergovernmental Panel on Climate Change (IPCC) Impact Assessments, organized by a global network of scientists hosted by the UN, which synthesize the available information on climate change every six or so years. Yet here Himalayan glaciers have fallen under the generic heading of 'High Mountain Areas', meaning they've been bundled up with the rest of the world's mountain ice – the Alps, Andes, African Tropics and so on.[12] It was only in 2019 that the first landmark impact assessment for the extended Hindu Kush Himalaya region (to include the Karakorum, Pamirs, Tien Shan and Tibetan Plateau) and its glaciers was written, led by the International Centre for Integrated Mountain Development (ICIMOD), a regional inter-governmental institute based in Nepal. It took five years to compile, covering topics including biodiversity, climate, energy, food security and water, synthesizing its diverse findings into a single, region-wide assessment.[13] The conclusion: the shrinking of glaciers has been widespread in the extended Himalaya since at least the 1970s. The glacier mass loss is even more pronounced as you move further east, as the Asian monsoon kicks in. For glaciers that receive snow from the monsoon, warming is especially bad news – mainly because rainfall replaces the snowfall which forms a protective, reflective blanket that slows melt.[14]

Some scientists have wondered whether the debris armour on Himalayan glaciers – which so hindered my passage up Chhota Shigri – could save these glaciers, by shielding them from more melt. Yet recent studies using satellite images seem to show that debris-covered glaciers are for the most part in no better health than clean glaciers,[15] partly because as all that debris builds up on the glacier surface, the ice eventually slows down; lakes then form on top of the glacier and large ice cliffs appear around its edges, both becoming hot spots for melting. All this extra melt undoes the good work done by the debris layer in slowing melt.[16] It's a confusing jumble of pluses and minuses for these debris-mantled glaciers, and for the scientists it's very much work in progress.

A few isolated spots, such as in the Karakorum, western Kun Lun (Tibetan Plateau) and the eastern Pamir, are bucking the overall shrinking trend.[17] Many of the Karakorum glaciers are large and at very high altitude, up to 8,000 metres above sea level, meaning they can receive snow all year round. In these places, cooler summer temperatures and higher snowfall from the persistent westerlies seem to be keeping the glaciers healthy for now, though the reasons are not fully understood. It's most likely that a combination of factors is at work, including the changing climate, causing shifts in the strength of the Asian Monsoon, alongside local factors such as the greater irrigation of the plains in western China, boosting moisture supply to the atmosphere and causing snow showers to fall on the mountains above. But the trend of glacier advance here seems unlikely to persist.

The Himalaya will almost certainly warm faster than the rest of the world (on average) over the coming century, simply because warming in the high mountains (just like in the Arctic) strips away reflective ice and snow surfaces that previously bounced the sun's rays back to space. So even if, by some miracle, nations manage to keep future globally averaged warming to the most ambitious limit advocated by the Paris Agreement – one and a half degrees Celsius above the pre-industrial level – the Himalaya could warm by around two degrees Celsius.[18] Best-case scenario: if we limit global warming to an average maximum of one and a half degrees Celsius, about a third of Himalayan glaciers might be lost by the end of the twenty-first century. More likely scenario: we continue burning fossil fuels at the present rate, and perhaps two-thirds of the glaciers will be gone by then.[19]

As for the great white glacial rivers that are the lifeblood of millions of people in the Himalayan region – how will they be affected? In the short term, until around about 2050, it's thought that glacier melt flows into these rivers will increase.[20] This is because the glaciers currently cover such a large area, and future warming will simply accelerate the melt of their snow and ice surfaces. But after the middle of the twenty-first century, the melt supply will fall as glaciers become too small to sustain large melt flows, even if their melt rate is still high.

Eventually, rivers will start to have considerably diminished flows at certain times of the year, impacting on domestic use, agriculture and energy potential via hydro-power operations – this is particularly true of the headwaters of rivers like the Indus, which are heavily reliant on glacier meltwater in the dry summer months. The white rivers will start to run dry. You might imagine that in the monsoonal rain belt, with reduced glacier melt, the rains would become proportionally more important – the problem here, though, is that the rain tends to fall in sharp, unpredictable bursts, rather than as a constant drip like glacier melt, meaning that local people don't know when the water tap will turn on and off.

The importance of glacial meltwater as a life source was never far from my mind at Chhota Shigri Base Camp; wherever I was, I could always hear the muffled thunder of the river as it charged past our camp, ultimately bound for the Indus. Yet beyond the narrow strip of this lifeline, the land was gasping for water, blanketed by sand, gravel and boulders organized into moraines and huge debris fans, or simply scattered messily in heaps. Water gives life, absence of water takes life away – this basic truth underpins just about every aspect of human existence here. It is for this reason that water and glaciers have such religious significance for the people of the Himalaya.

The Ganges is perhaps the world's holiest river, often referred to as *Ganga Ma* (Mother Ganges) by Hindus who consider her to be their mother goddess, and believe that her waters rinse away their sins. The river finds its source at the gaping ice cave at the snout of Gangotri Glacier in Uttarakhand, near the Indian border with Tibet – a sacred Hindu site where thousands flock every year to bathe in the holy waters. Not so long ago, however, the Indus was Hinduism's most revered river. Indeed, the word 'Indus' derives from the ancient Sanskrit *sindhu*, which simply means 'river' – giving rise to the names 'Hindu' and 'Hindustan', and then, via Ancient Greek, 'India'.[21] Since 1947, when India gained its independence and was partitioned, the Indus has largely flowed through Pakistan, and the water cult of India has moved to the Ganges.

This entwining of spirituality and water intrigued me. I was very taken by the idea that glaciers and springs were the realm of living goddesses with the power to cleanse and give life – for I had recently started to catch myself wondering, while trudging through the landscapes of Patagonia, whether there might be something more, something beyond what I could see, touch and sense? In these icy wastelands I had at times felt close to some kind of vitality that was neither human nor born of the terrain – a playfulness in the breeze as it ushered clouds up and over the soaring peaks, or the momentary warmth from the sun as it rose to quell dark, cold shadows, or (occasionally) an almost animate presence lurking at the fringes of a glacier. These were fleeting moments, lasting a mere millisecond, but long enough to spark a sense that, just maybe, there was a higher being at work.

As a scientist, I don't find this difficult to get my head round. There are a great many things in the natural world that science can't explain. It doesn't mean they don't happen. The same goes for events in my personal life. In the summer of 2013, a week after she died, I had an extraordinary conversation with my mother. Swamped by grief, I sat in the darkened room of a psychic medium, a sweet, thick smell of incense filling the air. A spirit being appeared – invisible to me, but not to the medium – and identified herself by my mother's name, accurately described her ailments, her feelings at the end, and even knew that I was wearing her wedding ring. I left feeling giddy at the thought that people might move between realms, slightly unnerved that my entire world view up until that point had been so very one-dimensional.

In Gilgit-Baltistan, a northern region of the Karakoram in Pakistan, many people see the glaciers as animate beings, and distinguish between male and female glaciers. The male ones are dark, move slowly and give little water (what scientists call 'debris-covered glaciers') whereas the female ones gleam white or blue and yield much water ('clean glaciers').[22] Locals have an ancient practice of mixing ice from the two, and placing the chunks together in a sheltered place, such as a mountain cave, interspersed with gourds of water, which eventually smash as their water expands during winter freezing, supplying water to grow the ice chunks. The mixed ice chunks are covered with charcoal, and/or other materials like branches or cloth, to insulate them and slow melting. In this way the female glacier is 'impregnated' by the male glacier, and over the following winters a new glacier starts to grow. Another method of propagating glaciers has now been commandeered for the dry, cold region of Ladakh in northern India, in a process initiated by the engineer Sonam Wangchuk, by which meltwater stored during the summer is in winter channelled down the mountain towards the valley, where its high pressure enables it to be sprayed upwards into the freezing air to create a bulbous icy pyramid called a 'stupa' – the name chosen to reflect the Buddhist stone ceremonial burial mounds of a similar form.[23] Positioned close to villages, these 'stupas' melt in springtime, before the glaciers, when there's often a water shortage for local communities. Such artificial glaciers must be grown again every year, but they do help bridge a gap in the water supply.

A similar aridity also typified the rocky terrain in front of Chhota Shigri. For something teetering so high up, our camp there was surprisingly sturdy, comprising a sterile-looking white Portakabin lined with bunks for sleeping and a solid stone hut set into the lumpy moraine behind, around which we pitched our tents. I had brought the same expensive tent that I'd previously lugged all the way to Patagonia, where it had created its own nocturnal sprinkling system of tiny droplets onto my head. Up here in the high Himalaya, this little hooped 'sarcophagus' (as one of my students, Matthew Marshall, dubbed it) was absolutely perfect. All problems of condensation vanished, and it became a cosy refuge against the bitter chill that swept the glacier forefields once the sun had slunk behind the mountains.

Even so, I barely slept at all at Chhota Shigri over the week or so I was there. I was constantly aware of my need to breathe deeply to draw in sufficient oxygen, and was plagued by a dull ache at the back of my head. When the dark canvas started turning grey with the approach of dawn, a deep sense of relief washed over me. I'd crawl out of my long canvas tube into its slightly larger vestibule and hurriedly layer on as many clothes as possible – this was definitely 'two down jackets' territory. I'd normally first wander over to check on Jon Telling, who had decided to defy tents altogether, installing himself about ten metres away from my sarcophagus in a long crevice beneath the overhanging flanks of a large boulder, kept warm by his khaki camouflage Latvian Army bivvy bag, which he swore was better than any canvas – although he did admit one morning that he'd felt a bit nippy the previous night. Both of us bettered Pete Nienow's sleeping arrangements, though – he'd borrowed a tent from our Indian friends which, having weathered many seasons at Chhota, finally succumbed to high winds one stormy night and collapsed dramatically onto his nose. The Indian contingent wisely opted for the Portakabin and were usually already up and about by the time we emerged, and one of the smiling Nepali guides always brought over a steaming tin cup of delicious, sweet lemon tea. I'd eat my breakfast perching on a boulder as the sun gradually bathed the rocky terrain in a soft, golden blush.

It seemed that our Indian colleagues were worried about

catering for us – they certainly made an enormous effort to conjure up all manner of Western breakfast edibles, from pancakes to porridge to egg baps, and took some persuading that we were perfectly happy with dal and chapatis. (We were used to preparing our own food in field camps, so having this decadent degree of service felt very unusual.) They were also touchingly concerned for the comfort of the women in their midst, and had hauled up the mountain an entire ceramic WC, complete with cistern and flush, which was installed in a small tent downstream from the camp. I'd never seen anything like it.

One quite remarkable thing that I stumbled upon while ambling around camp early one morning was a small, fenced enclosure, a few metres across, from which sprouted a profusion of green beans. The Nepali camp manager, Adhikari, explained that he'd planted the beans to see whether they might grow, if given a helping hand in the form of sheep and goat droppings. This set my mind whirring. These beans were growing on glacial sediments, once part of a solid mass of rock beneath Chhota Shigri. I recalled that the glacial flour which was released from the Greenland Ice Sheet had contained elements like phosphorus and potassium that might sustain tiny plant-like organisms floating in the sunlit surface ocean. Here I was, high up in the mountains, far from the sea, and yet the flour from Chhota Shigri seemed to provide a nutritious medium for these green beans against a backdrop of rubble and boulders. I wondered whether, if I took some of this glacial flour home, I might grow crops in it?

A year later, with the help of Sarah Tingey, who'd developed a rare passion for both plants and glaciers, we'd taken over a greenhouse on top of the University of Bristol Life Sciences building. Very soon, hundreds of soy bean plants were vigorously sprouting in pots filled with a mix of sand, which has almost no nutritious value on its own, and a gram or two of Chhota Shigri glacial flour. (Glaciology meets plant science, who'd have thought it?) It seemed that the flour assisted the growth of the soy beans just as well as a traditional chemical fertilizer, which has the downside of causing the slow degeneration of farmlands and pollution of groundwaters and streams. Plants generally grow well on land where there's soil organic matter; but at some point extra nutrients will probably be needed to stimulate growth, especially if one keeps on harvesting the plants for food, rather than letting them die and break down to form that precious organic matter.

Glacial flour can provide some of those nutrients that come from rock. The problem is that rock doesn't contain much nitrogen. It's the reason glacial meltwaters caused a problem for the phytoplankton in Patagonian and Greenland fjords – not enough nitrogen. However, if you take a crop that doesn't need nitrogen, such as a legume like the soy bean that can cleverly fix its own nitrogen, and add water plus a tiny amount of organic matter, such as goat droppings, the plant will flourish. I wondered whether glacial flour could be used to fertilize degraded croplands in poor rural farming communities in the Himalaya.

To extend this notion further – if glaciers are retreating, will they expose new land surfaces for plants, and even crops, to be nourished by glacial flour? In fact, when you look at the land in front of glaciers, there are already clues that this is happening. Near any glacier, you find heaps of freshly milled sediments, stuff that not long ago was hidden deep beneath the ice. Little can grow here – there's not enough organic matter and nitrogen. Yet moving away from the glacier, into older glacial flours, things start to happen. The hardier forms of life, such as microbes, lichens and mosses, move in and start to grow, slowly building up organic matter in the soil as they pass through the cycle of life and death. Some microbes take nitrogen out of the atmosphere to form parts of their cells; when they die, this gets added to the soil.[24] Slowly, over time, small plants move in, then large plants, then shrubs, then trees. This is called natural 'succession' – one form of ecosystem replaces the other as a mature ecosystem develops.

This natural colonization of new land surfaces is happening all the time in the seemingly barren landscapes in front of glaciers, and will increasingly occur as the snouts of glaciers further retreat. Analysis of satellite images taken of Nepal and its mountains already shows a 'greening' of the Himalaya in the belt of land which is sandwiched between the treeline and the snowline above it.[25] These changes since 1993 – around the time I became a glaciologist – suggest that plants are starting to grow where previously it was too cold, and presumably tapping into sources of nutrients like glacial flour. And if plants are growing naturally at higher elevations, then why could this not also happen artificially for crops like beans? It seems increasingly possible – over roughly the last fifty years, climate warming has extended the length of the growing season by around four days per decade across the Himalaya.[26]

However, there remains a big water-supply problem. As we've seen, the glacial meltwater 'tap' which feeds the rivers in this region will weaken later this century, reducing stream flows and making them more unpredictable. In a bold attempt to pre-empt this looming crisis, the governments of Asian nations have produced ambitious plans to dam major rivers to store water and help generate power. India has launched its grand 'National River Linking' project, which will connect up forty-four rivers across the subcontinent, including in the Himalaya, with 9,600 kilometres of canals.[27] The advantage of dams and reservoirs is that they act like a plug in the sink – even if the taps of glacial melt and rain reduce to a trickle, water can be trapped for use in arid times.

Unfortunately – as in Patagonia – these vast artificial lakes also trap sediments (such as glacial flour) which are so vital for sustaining agriculture on the flood plains of the world's rivers. Entire civilizations have been built on the green currency of the rich alluvial sediments borne by rivers – Ancient Egypt being a good example.[28] These days the Nile carries almost no sediment, due to large dam projects – the Pharaohs would struggle today.[29] It's possible you could take sediments accumulated in these dams for use further downstream – but you have to ask whether it would have been better not to have built the dam in the first place? The Himalayas are greening, and perhaps new areas will open up for agriculture, but solving the problem of water will be no mean feat. Glacier melting is one of the big humanitarian time bombs of climate change which will affect highly vulnerable communities – ironically none of whom are emitting much carbon dioxide.

On our final day at Chhota Shigri, before returning to the plains of Delhi, we once again lugged our cumbersome collection of crates and rucksacks down to the Chandra River crossing. My heart felt leaden, almost as if I were dealing with some form of loss. Hanging back until the end of the line, I was the final person to crawl into the clunky metal box. This time around, I simply relaxed into the motion, enjoying the brief moment of anticipation as the crate tipped off its rocky ledge to swing like a pendulum above the foaming white waters beneath, before being dragged in slow, jerky motions to the other side of the gorge. All this time, I never took my eyes off the frayed summits, trying to etch their precise form into my memory – the silhouettes of the peaks, the patchwork of light and shade, the smatterings of snow that lingered in gullies, all before they were taken away. Shortly after I reached the other side, Pete said to me, mildly bemused: 'Jems, did you realize that you were the *only* person who travelled across the river backwards, still facing the glacier? Everyone else faced the direction they were going!' I hadn't noticed, because I was too busy saying farewell.

# 7. The Last Ice

## The Cordillera Blanca

My trip to the Cordillera Blanca was the expedition that almost never happened. This was not just my first glacier expedition in Peru, but my first glacier following emergency brain surgery. The headaches which had started in India had worsened by the time I arrived in Patagonia in October 2018, but still I'd ignored them, and soon they were accompanied by fainting fits, numbness in my legs and the deterioration of my eyesight. Shortly before Christmas I'd been rushed into hospital to discover that my brain was being squeezed by a satsuma-sized 'benign' growth that was on course to kill me.

A downside to my ability to cope with extreme discomfort during expeditions was that I'd been able to plod on. One night at home I'd walked across the landing, passed out and fallen backwards through a large pane of glass – a picture I'd recently had framed that was idly propped against the wall. I woke up thinking, *time for breakfast*, but wondered why I couldn't move my head, which was now framed by a sheet of jagged, broken glass, its sharp edges cutting into my flesh, rivulets of blood trickling down my neck.

At work, I'd developed a way of camouflaging my symptoms. When I stood up at the end of meetings, I'd turn to gaze out of the window, pretending to admire the view (normally of the Bristol drizzle) because I could barely see, could not feel my legs, and my ears were filled with deafening sirens. Yet

I had no intention of seeing a doctor – I'm not sure why. Hassle, maybe, or fear, or not wanting to let others down at work? Or perhaps because my symptoms were reminiscent of those of my late mother, who had died after her cancer advanced to her brain? I honestly don't know.

Surgery was a swift affair. One day I was in a departmental meeting, the next I was out cold, face down on an operating table with my skull being sliced open. The best thing about my brain 'hiccup' was that it forced me to take time off work, and with that came plenty of space to think. Still, my convalescence was marked by some bleak moments as I wandered through Bristol with Poppy, my beloved Labrador, in the drab winter greyness. The euphoria of having survived brain surgery was soon replaced by waves of gut-wrenching emotion as the trauma worked its way through me, and I grappled with the meaning of existence – all that I had ever relied upon to make myself feel good or to validate myself was removed, slowly and painfully, like layer upon layer of paint stripped off an old wall. Who am I? What do I believe in? It felt like I was being exfoliated back to the real me, to the person I was before ambition crept in, before I stepped onto the conveyor belt of life and work.

Fortunately, at a time when I lacked belief in almost everything I'd ever done or would ever do, my passion for glaciers remained undimmed, and I tentatively began to wonder how I might share my discoveries as widely as possible, while I was still alive. I started to make plans again. I found a quirky gym around the corner from my house – it was just what I needed, frequented by an eclectic bunch of friendly souls, including a number of jovial pensioners who turned up in their corduroy trousers and crisply ironed shirts, determined to turn back the clock of muscle decline by pumping iron. I began lifting weights daily, particularly to strengthen the severed neck muscles where the surgeons had cut around my skull, and gradually I got stronger.

Mentally, my recovery was slower. Cognitive tests put me within the bottom 2 per cent of the population for some aspects of my memory performance – not a great situation for a Professor of Glaciers. I bungled my words in conversation. Most terrifyingly, I could no longer read a map. These impairments extended to scientific writing – the required precision and depth now seemed inaccessible to me. Then one day I tried a different tack and started for the first time to write creatively about my life with glaciers. I had no expectations of what I could or couldn't do – it was my own secret project. It felt very liberating.

Despite not being quite one hundred per cent, eight months after my operation I'd decided that I needed to get back on a glacier. My team in Bristol had been busy planning the trip to Peru for many months, led by a postdoctoral researcher, Moya Macdonald, who'd recently finished her PhD on Svalbard and was eager to hop on a project which involved ice and people (for a change). A woman of fiery determination, Moya had a brain that can juggle a phenomenal number of complex logistical matters, while also intelligently contemplating a science problem – without her I would not have had an expedition to join. We'd written the winning grant proposal to examine the precariously positioned glaciers of the Cordillera Blanca more than a year earlier, and had been looking forward to the project ever since – however, brain surgery had kind of got in the way.

This isn't to say that I wasn't terrified by the prospect of joining an expedition again. I was still emotionally drained, and friends stoked my misgivings by casually asking, 'Jem, are you sure Peru is a good idea?' and 'What do the doctors think?'

(Good question – I hadn't asked them!) I didn't think that I would be fit to lead an expedition again after what had happened. What if I couldn't solve problems like I used to? What if I couldn't work long days? What if people noticed that I was struggling? What if my head couldn't cope with 5,000 metres of altitude? What if my balance was so bad I toppled off a rock, into a raging torrent, and drowned? So many what ifs! Weird though this may sound, I often imagined a reptile on my shoulder, constantly chuntering away, amplifying my worst fears. But somewhere deep inside, I found the willpower to silence that scaly beast.

Moya and I landed in Lima one sullen, overcast day in July 2019, to be reunited with our new collaborator Raúl Loayza-Muro and his assistant Fiorella La Matta. I'd first met Raúl by chance at a workshop about Peruvian glaciers in Lima in March 2018; we'd ended up on a table together – two lonely water chemists, and used the time to share field stories and hatch plans for a project to investigate the impact of Peruvian glacier retreat on river-water quality. Raúl was one of the most laid-back scientists I'd ever encountered, and he soon endeared himself to us with his potent pisco sours and his capacity to make us laugh in just about any situation.

Our journey to the Cordillera Blanca began with an eight-hour skeleton-rattling drive to the north. Leaving Lima via its congested highways felt like repeatedly trying to evade the jaws of a giant beast that occasionally let go, then snapped shut again just as you were about to run free. It was interminable, and I honestly believed we were never going anywhere. But finally, after a couple of hours, we reached the city limits and were greeted by the lifeless sands of the coastal desert to the right, the ocean to the left. Rolling yellow dunes, softly greened in areas moistened by the sea mist, graded to desiccated mountains rising from the flat dusty plains, their flanks sculpted by run-off during the rainy season. The view was blurred by the suffocating haze of yellow and grey dust. Although this was my first 'proper' desert, it reminded me of the barren terrain around glaciers – another habitat where life struggles to survive in freezing temperatures with only an intermittent water supply.

After about four hours we turned off the coastal highway and began the slow climb into the Cordillera. Seen from space, this lofty mountain belt strikes an impressive white gash running north-north-west / south-south-east across northern Peru, following a pronounced fault line that separates the Cordillera Blanca to the east (meaning white range, because of its ice) and the Cordillera Negra to the west (black range, with no ice). It was dusk by the time we neared the range, the sun dropping out of the sky at around 6 p.m. at this tropical latitude. Cramped in the back of the 4 x 4, struggling to keep the contents of my stomach down as we wound up the twisting road, I could just make out the distant iced peaks of the southern Cordillera, peering down on the serene flatness of Lake Conococha. (Cradled by mountains, its name is thought to originate from the Quechua words *cúnoc* for warm and *cocha* for lake, due to the hot springs that emerge along its western banks.) Although it was beautiful, I was shocked. The glaciers which clung to the mountains here were the tiniest ice forms I had ever seen; they had retreated to the peaks, where their stumpy, sickly tongues glowed pink in the sunset.

Glaciers in the tropics are particularly sensitive to a warming climate – they are mostly small, at high elevation, and their health depends on keeping the freezing level as far down the mountains as possible.[1] In Peru, the Little Ice Age peaked in the mid to late 1600s, but since then the climate in the Andes has been generally warming; during the second half of the twentieth century this warming was up to 0.3 degrees Celsius per decade, about five times the globally averaged rate.[2] The amount of snow falling on the region's glaciers has slightly increased over the last thirty years, but not enough to counter-balance the additional melting. As a result, over the past two decades the glaciers of the Cordillera have shrunk by a stagger-ing 30 per cent.[3] Computer models of the local climate predict that all but the very tiniest blobs of ice on the highest moun-tain tops will have disappeared by the end of the twenty-first century if we do nothing about our greenhouse gas emissions.[4] If, on the other hand, we take decisive, collective action to cut carbon emissions such that they fall to zero by 2100, then the same models show the outlook is much more positive. The glaciers will continue to shrink, but perhaps about half of those in the Cordillera Blanca will not vanish after all.

Despite their shrinkage to tiny stubs in many places, these glaciers are a vital source of water to local communities. In this tropical climate, air temperatures do not vary hugely over the course of a year, but there's a dry season and a wet season – some 70 to 80 per cent of precipitation falls between October and April.[5] During the dry season the water supply must come from storage somewhere, and glaciers are like enormous fro-zen reservoirs. These Andean glaciers receive snowfall in their upper basins mostly during the rainy season, but because this is the tropics and it's warm all year round, their lower tongues melt during both the wet and dry seasons – unlike the Euro-pean Alps, where the glaciers are protected for half the year. (The alpine winter is too cold to allow for much melting; instead the whole glacier plumbing system simply shuts down.) The problem in Peru is that over the past few decades the gla-ciers have become so small that they are on what one might call the 'declining curve of water availability'. Elsewhere in the world, such as Patagonia and the Himalaya, meltwater supply from many glaciers is still rising as melt rates increase – but if they continue shrinking, of course, their melt rates will ultimately dwindle too.[6]

The Cordillera Blanca is the largest ice-topped tropical mountain range, home to about a quarter of the world's tropical glaciers – after spending twenty years hanging out at the poles, or not far off, the concept of a 'tropical glacier' intrigued me. Moreover, the meltwaters of these high Andean glaciers presented a massive conundrum. The rivers they fed had the strangest of chemistries – like nothing I'd ever encountered before. Many of them had become highly acidic, and I mean *crazily* acidic, almost as much so as your stomach or lemon juice (about pH 2–3), and what's more, they were intoxicated with high concentrations of heavy metals such as arsenic and lead. Definitely *not* drinkable. But what was causing such an extreme chemistry? Many scientists were linking the toxicity to the retreat of glaciers, but I couldn't for the life of me figure out how this might be so. Rivers emerging from every other glacier I'd worked on were more or less neutral to alkaline in their pH, regardless of what rocks they ran over – which made them pretty good for drinking (after filtering out any glacial flour). I soon realized that it would be almost impossible to solve this mystery without delving into how the Cordillera Blanca and its glaciers came about in the first place; essentially, I needed to go back in time.

The very existence of the Cordillera Blanca has puzzled many geologists, who today still debate how it formed.[7] At the heart of the enigma is the fact that this range straddles two separate chunks of the Earth's crust, the South American Plate (which includes the continent) in the east and the Nazca Plate (which sits mostly over the ocean) in the west. These tectonic plates have slowly been converging for at least the last twenty million years, the Nazca Plate plunging ('subducting') beneath the South American Plate just off the coast – this convergence causing the formation of the Andes. Yet we now know that the Cordillera Blanca developed just five million years ago, in a setting where there's local stretching apart of the Earth's crust via what we call a 'normal fault', which exists when one side of a rupture in the Earth's crust slips down, leaving the other side raised above. As a result of this, on the west side the Cordillera Negra has dropped down to make a lower step, and on the east side the Cordillera Blanca sits as a raised foot towering over its darker neighbour.

These two mountain belts are chalk and (cinder-coated) cheese. To the east, the heavily glaciated peaks of the Cordillera Blanca are nourished by a few metres of snowfall every year, fresh from the Amazon Basin.[8] To the west, sitting as it does in the rain shadow of the Cordillera Blanca, the Cordillera Negra receives insufficient moisture to form glaciers. The Río Santa, a vital source of water to the people of this region, runs between these two strikingly different mountain ranges, very close to the fault line. As the river travels from south to north, it collects melt from many glaciers, incising ever deeper into its underlying rock as it carves its way to the Pacific.

To understand why the glacial rivers of this region might be so poisonous, it's necessary to wrap one's head around its mind-boggling geology. Sometime between thirteen and five million years ago a mass of molten rock (magma) was ejected up from deep within the Earth, cooling to form a huge granite pluton or batholith[9] (a dense mass of solidified magma) that was emplaced into much older sedimentary rocks known as the Chicama Formation, which had been transformed from marine deposits during the Jurassic Period about 150 million years ago. This upward-moving magma helped push up the Cordillera Blanca, and the batholith now sits within the raised footwall of the normal fault, with a coating of soft Chicama rocks, topped off by glaciers at five to six thousand metres above sea level – in other words, a kind of three-layered cake of hard rock, soft rock and frozen icing.

However, as the glaciers retreat, the Chicama rocks, which are full of metal-rich minerals such as sulphides and ores, are slowly becoming exposed to the air. One particular mineral, iron sulphide or pyrite (fool's gold), where one iron atom is bound up with two sulphur atoms, $FeS_2$, is found here at notably high levels – much more so than at any other glaciers I've known. Pyrite is an extremely reactive mineral, interacting with oxygen in the air to produce sulphuric acid and oxidized forms of iron, like what we know as rust. This means double trouble for water quality – for not only is highly acidic water on its own not good for human consumption, but the acidity also increases the solubility of toxic metals like arsenic, lead and mercury.

Splashed across the flanks of mountains high up in the Cordillera are red and orange streaks of the Chicama rocks – testament to the high concentrations of iron held within, which form rust upon exposure to the air. Mining companies are active in the region, extracting valuable metals such as lead and copper from these rocks. Unfortunately, when you mine the landscape and crush it up into fine particles, as soon as any metal sulphides are exposed to air and water they dissolve rapidly, poisoning lakes and rivers – a phenomenon widely known as 'Acid Mine Drainage', due to its association with this grubby business. Minerals account for more than 60 per cent of Peru's exports,[10] and mining has caused severe tensions to flare up between local communities, national park managers and mine owners. Even scientists have been caught up in this complex web. A team of researchers were kidnapped in August 2019, around the time we were there, by villagers who suspected they intended to exploit glaciers for minerals.

Glaciers are, in their own way, natural mines, crushing and grinding the rock beneath them as they move. It's been suggested that as glaciers retreat they leave their mined waste in their forefields, and any exposed metal sulphides will oxidize rapidly to cause the acidification of waters and metal toxicity – raising the question, how will future glacier retreat impact the water supply? How long before the region's rivers are undrinkable?

The purpose of our expedition was to find out what was causing the toxicity in river waters and, if possible, come up with a solution. We started off during the dry season, so that we could enjoy the benefits of the clement weather while working out the lie of the land; much of this time was spent in a pickup truck on steep dirt tracks etched into the mountains, following major rivers from their glacial source to their meeting with the Río Santa. As we combed the valleys of the Cordillera, from the wide sweeping grassland valleys of the far south to the deep-cut rocky ravines of the north, I felt I was back in my element – the heady thrill of my undergraduate days in Arolla rekindled for the first time in years, as I leapt over boulders and clambered down gorges, my pH probe ever ready to discern the acidity of the water. After six months off work, my days filled with dog walking and tea drinking, this was a big step up in pace for me.

Every morning, at first light, we would depart the buzz of Huaraz – location of the infamous Glacier Lake Outburst Flood disaster of 1941 – and head for the hills, with the aim of covering several river catchments that day. Being for once based in a city, we were staying in a hotel – a modest one, mind you, but still luxurious compared to our usual remote field camps under canvas. Unfortunately the urban setting meant that we couldn't leave any equipment in the truck at night in case of theft, which necessitated a good half hour spent hauling a few hundred kilos of gear, including a freezer, up and down several flights of stairs every morning and evening. The early mornings had a sharp bite to them, which always sparked a little exhilaration and quickening of the senses before a long day of adventure; by mid-morning, though, as the sun climbed high into the sky, the air was suffused with warmth. Overall, we worked on about thirty rivers during our ramblings, and something that greatly puzzled me was why some of them were acidic and metal toxic, whereas others were not; also, how did some start off acidic in their headwaters, but manage to recover prior to evacuating their water into the Río Santa? It would take a long time to tell the story of all the rivers we examined – so I've chosen to tell the story of just two of them.

The first of our duo is the Pachacoto River in the southern Cordillera. An energetic torrent in its lower reaches, even during the dry season, it captures melt from several glaciers that hug the high peaks, including the Nevado Pastoruri. This is perhaps the Cordillera's most famous glacier due to its accessibility to tourists, who arrive by the busload daily, staggering along the trail to the glacier at over 5,000 metres above sea level, gasping for air like fish out of water, to finally gaze in wonder at its steep, gleaming, icicle-encrusted snout. Between 1975 and 2010 this ill-fated glacier lost around five square kilometres of its clean ice area, roughly halving the size of its dome.[11] Such rapid melting has led to the separation of the main body of Pastoruri into an eastern and a western mass, with a few 'remnant blobs' (to borrow a new technical term invented by Moya) dotted to the sides. I'd heard numerous reports of its meltwaters becoming highly acidic and toxic as the glacier has retreated, exposing mounds of debris in its forefields – so I was confused when, at the confluence of the Pachacoto River and the Río Santa some twenty kilometres downstream from the glacier, we found the waters to be utterly ordinary and pleasantly alkaline. How could this be so?

As we drove up the dirt road that traverses the tops of for-mer river terraces in Pachacoto's valley, I admired the glistening peaks, soft yellow grasslands, and clusters of strange cactus-like trees, ten metres tall, resembling aliens just landed from a far-off planet – these were *Puya raimondii*, a member of the Bromeliad family and cousin of the pineapple that is native to the uplands of Peru and Bolivia. (The puya can take as long as eighty years to flower, whereupon it releases millions of seeds into the air, shortly afterwards withering and dying, its job done.) But halfway up the valley, as we rounded a bend, I jolted bolt upright – for suddenly I caught sight of the snaking river, and realized it had turned orange.

From the road, it looked like an interloper, out of place in this seemingly pristine high Andean paradise. But upon closer inspection, following a tiring trudge, I was speechless to dis-cover that the gushing mass of clear river water was fringed on all sides by a bright orange bed and banks. I turned over a stone, only to find nothing living beneath – just a gungy red algal mass and thick rust.[12] Sitting on a boulder to wonder at this spec-tacle, I too acquired an unpleasant bloom of russet powder. So here were the first signs of toxicity in the Pachacoto. Confus-ingly, though, its waters were not acidic – about pH 7. Clearly something had neutralized it, which must explain why metals like iron were fast dropping out of solution. So where had this highly metal-toxic water come from? And how was the river managing to recover before it entered the Río Santa?

My first question was answered when we finally arrived at the Pastoruri Glacier at the head of the valley. As I shuffled along the trail in time with the convoy of tourists, I couldn't help but reflect, rather gloomily, that it felt like a funeral procession to witness the dying days of this esteemed glacier. This was certainly a far cry from most of my icy excursions, which are typically to wild, inhospitable places where few people go. My head was pounding in the area where I'd had surgery seven months earlier. Since I'd done relatively little exercise while recovering, and certainly wasn't mountain-fit, I was also battling hard to fill my lungs with sufficient air. Strangely, though, despite the discomfort, my legs powered me onwards – Moya jokingly referred to this as having my 'ice legs' on.

The front of Pastoruri is steep and imposing – gleaming white cliffs more than ten metres high, overhanging in places. The surrounding terrain glared an angry orange, indicating the exposure of the Chicama rocks – once secreted beneath the glacier, now bared to the elements. I dipped my pH probe into the tiny stream draining out from the front of an ice cave sculpted in the glacier snout; it registered pH 2. To my surprise a tourist came over to ask us to refill his water bottle, and we broke it to him that perhaps this wasn't a good idea – it would be a bit like drinking a litre of lemon juice infused with toxic metals. Such staggering acidity seemed almost incredible, for my research over the years had taught me that as glaciers crush the rocks beneath them, they liberate acid-producing minerals like sulphides, but they also release carbonate minerals from rocks such as limestone, which consume acidity. Here at Pastoruri there seemed to be way too much sulphide in the Chicama rocks, and not nearly enough carbonate.

And yet, rather astoundingly, some twenty kilometres downstream, by the time the Pachacoto River converged with the Río Santa, this toxicity had attenuated and metal concentrations had fallen below harmful levels.[13] Clearly, in remediating the waters, there were natural processes at play. It turned out, while driving up and down the valley, that it was not difficult to work out what these were. The Pachacoto valley, like others in the southern Cordillera, is very wide, bounded on both sides by sweeping glacier moraines formed of non-toxic granitic rocks from the batholith. Along its gentle borders you see springs popping up out of the hillside where there's a break of slope, supplied with groundwaters moving down inside the mountain from high pressure to low pressure – and all these springs are alkaline in pH, with low metal concentrations. Thus the gradual supply of these pristine waters to the toxic waters of the main Pachacoto River move it towards healthier conditions, and, as the acidity falls, so metals start to drop out of solution to form the blanket of orange rust I'd seen halfway up the valley. The wetlands of this flat-bottomed valley – large expanses of boggy ground held together by mosses and grasses – also play a role in reducing the toxicity of the Pachacoto River. Locally known as 'bofedales' (or pampas), many of the plants are native to the Andes, with an ability to take up large amounts of metals directly from river water. The power of this natural water filter is something that local people are now actively trying to harness to make the rivers drinkable again.

Sadly, it's a very different tale in some of the more narrow, steep-sided river valleys of the northern Cordillera. When I visited Shallap Glacier for the first time, right at the end of the expedition, I was not having the easiest of days. Long days on the road, too much excitement and too little sleep – I'd hit the wall in terms of what my body could cope with. My head was sore and my legs were leaden; every time I stood up I teetered for a few seconds, hoping not to pass out. However, Shallap was a glacier that I was desperate to see and to sample – so off we set up the dirt track to the glacier, first in a 4 x 4, then on foot.

More than 5,000 metres above sea level, Shallap Glacier flows down very steep rocky terrain; large amounts of snowfall on the upper glacier combined with high melting on its lower flanks mean that speedy ice flow is needed in order to redistribute its mass. Like most glaciers in the Cordillera Blanca, Shallap presents a heavily crevassed, stumpy tongue, partially blanketed by debris from slopes above. This tongue, which is considerably lower in altitude than the main ice body, melts rapidly and renders the entire glacier very sensitive to warming – as it will continue to do until the tongue disappears and it can retreat to high ground.[14]

Like Pastoruri Glacier, Shallap is retreating over metal-rich Chicama rocks, clearly evident from the rusty red-stained landscape before its snout. Again, the juxtaposition of the striking beauty of the steep rock-walled valley and its otherworldly orange-lined river made my mind do somersaults. A lake lay with eerie stillness in the bedrock depression carved by the glacier when it filled its upper valley – murky green from the very high concentrations of toxic minerals dissolved in its waters. The valley is a popular hiking route, and guidebooks refer to the lake's 'exquisite green waters' – refraining from spoiling the idyll by mentioning that it is poisonous.

Even at the bottom of Shallap valley, eight kilometres downstream from the glacier, the river waters remained highly acidic and toxic with metals, the river bright orange. This was a very different story to Pastoruri – but why? In Shallap the boundary between the metal-rich Chicama rocks and the granite of the batholith is clearly visible about a third of the way down the valley from the glacier. The Chicama generates acidic waters, but mixing with more alkaline waters flowing over the batholith ought to mitigate the acidity. The problem is, though, that the valley lacks large broad hillsides and moraines to soak up rainwater which forms neutralizing groundwater reserves; the only places where rainwater can be stored are in small, steep talus cones (from rockfall) which line the steep valley sides, or in small lakes or snow patches high up above. Tiny streams and waterfalls plunge daringly down the mountainside and into the main Shallap river, but (unlike the springs in Pastoruri valley) they are neutral to slightly acidic in their pH. Acid-consuming wetland pampas, which can absorb metals from waters, only cover a tiny area on the valley bottom. Essentially, the Shallap River cannot recover from the slug of acid input from its feeding glacier above. The concern is that many other valleys like Shallap will head this way, and the all-important local water supply will become poisoned.

You start to sense the potential scale of this problem as you travel through the region. Farms and small settlements of the native Quechua people, whose language is spoken by a sizeable minority of the population of Peru, are dotted up the hillsides to seemingly impossible heights in many valleys, even where rivers are a tell-tale orange. These sombre-faced people make use of every tiny fragment of land, growing crops in rotation on lower land and grazing sheep, alpacas and cows higher up; the women are colourfully dressed, even on the most ordinary of days, and are often seen sporting tall black hats, the height of which can denote their village of origin. Only a few decades ago the Quechua could fish in many of the river valleys of the Cordillera, which were well stocked with trout and other edible species; yet now a number of rivers are devoid of life. These high Andean communities, in many ways, are most vulnerable to changes in glacier melting, for without a reliable, clean water supply, they cannot thrive at these harsh altitudes.

My journey to the very northerly tip of the Cordillera, a landscape which is heavily mined for its ores, had painted a corresponding picture. Here the rivers were an assortment of shades, from orange to auburn, and intensely metal-toxic. There are no glaciers here – this is purely a side effect of the presence of metal-rich rocks, worsened by the mining – and yet it's remarkably similar to valleys like Shallap, reinforcing the notion that in some ways glaciers act as natural mines. Across the region, where mining or glaciers have led to acidity and toxicity, people are actively searching out non-contaminated water. The Quilcay River, which supplies Huaraz, has been blighted as the glaciers have retreated; thus the inhabitants of this regional capital now drink its flow mixed with water from more pristine sources.[15]

If more rivers become toxic in the Cordillera Blanca, it's unlikely that this situation can be reversed. Glacier retreat will continue – although the extent of this clearly depends on what we do about our global greenhouse gas emissions. Meanwhile, at a local level people are starting to adapt and find resilience, following in the footsteps of the Incas, who were experts when it came to water engineering and building aqueducts. The community across the Shallap catchment recently constructed a concrete channel that conveys water from the river for twenty kilometres out of its valley, closer to the settlements where it is needed. To remedy the problem of the water's acidity and toxicity, it is funnelled through an artificial wetland high up on the hillside. First, the water is directed over a series of small waterfalls, to be aerated and to stimulate the oxidation and precipitation of metals. Next, to counteract the acidity, it flows through a bed of packed lime, also causing more loss of metals from the water as they precipitate in more alkaline water. Once that's done, the water is directed through a convoluted network of wider channels containing metal-tolerant plants that filter out further metal toxins. The resulting clean water is finally supplied to villages, and supports more than 3,000 families.

Many of the Quechua of the Cordillera Blanca have a deep spiritual relationship with their water and glaciers through the spirit gods – the *Apus* (Lords) and *Pachamama* (Mother Earth) – who reside in the mountains and ensure the success of their crops. They offer sacrifices to the gods before planting their crops, symbolically feeding the soil with coca leaves and alcohol, praying for a strong harvest. Further south, in the highlands nearer to Cusco, pilgrims travel from across the region to the *Quyllurit'i* Festival, which has both Christian and indigenous Andean roots. *Ukukus*, men who mediate between the mountain spirits and the people, trudge up onto the glaciers; until recently, they came back carrying enormous chunks of ice, but this is no longer allowed now that the glaciers are fast disappearing.[16]

High up in the Cordillera Blanca, I felt a renewed sense of life and wonder. My recent close encounter with death had left me tussling with many questions – the main one being, how come I was still alive? In the past, I'd diced with polar bears, deep crevasses, raging rivers . . . and now I seemed to have survived brain surgery. When I visited Pastoruri Glacier – my first day back on ice after my illness – I felt a tectonic shift in my mind. As I moved towards the ice, having ducked under the barriers keeping the tourists at bay, the buzzing of human voices dropped away and I was left with the music of trickling meltwater as it dropped down icicles hanging over the cathedral-like white cliffs that towered above me. This colossal mass of terminal ice had been transported slowly, by flow, from its birthplace in the icefields, to meet its liquid fate. As I approached, tears streamed down my face – this glacier was so beautiful, so solid, so pure and yet inexorably melting away. I leaned in, stretching out my arms to embrace it like an old friend. I pressed my face into the tiny sharp ice crystals on its vertical face, and its melt combined with my tears, and together they flowed down my face. Maybe it would be here in twenty years, maybe it wouldn't. Maybe I would, maybe I wouldn't.

Before flying back to Bristol at the end of my trip, I met up with the Peruvian actress and storyteller Erika Stockholm. We'd been introduced through a programme run by the Hay Festival and the Natural Environment Research Council of the UK, which connects scientists with artists in order to find new ways to tell the stories of our research. Huddled in a Lima café on a grey, dank day, we discussed what it meant to inhabit our seemingly polar-opposite worlds. The superficial differences between us were obvious – I turned up in frayed jeans and a scruffy fleece, hair dishevelled from a sweaty hike across the city. Erika, on the other hand, was immaculate – jet black hair stylishly swept over her striking angular face, eyebrows carefully coiffured, lashes dramatized by a thick layer of mascara. The scientist meets the artist. Yet our conversation, although initiated in the usual faltering style of two strangers, rapidly soared into a passionate sharing of ideas and experiences – it was one of those rare occasions where you develop a strange sense that you've known someone for a lifetime.

We learnt that we were both storytellers, but constrained by different conventions and notions of how our stories should be told. Mine in data and facts; hers in happenings and feelings. We started, over coffee, to craft a story about Shallap Glacier and its dramatic tale of toxicity. Over the following months,

Erika brought our fledgling narrative to life. Together with my friend and collaborator Raúl Loayza-Muro, we'd perform it at the Hay Festival, to be held during November 2019 in Arequipa, Peru's second city; the three of us would take different roles in the story. During the course of many Skype conversations between Bristol and Lima, what crystallized in me was a desire to express Shallap's story in the most personal way possible. At one point, without thinking, I blurted out the words, 'I could become the glacier!' And so it came to pass.

Breaking with a lifetime of scientific rigidity, I painted my face white and donned a flamboyant blue wig, then cried, shouted, danced and moved as if I were indeed the dying glacier – in front of more than 150 people. I'm an experienced public speaker, but during the run-up to this particular engagement I'd been so petrified that I couldn't even bring myself to look at the script – every time I opened up the document on my computer, I'd feel a sharp flutter of panic and quickly snap the laptop shut. It wasn't until two weeks before the performance, while I was in Patagonia, that I realized that if I didn't conquer my fear, it was going to be an utter disaster. I crouched in my tiny tent, surrounded by the gentle tapping of raindrops on canvas, and proceeded to recite my lines, each word creating a trail of mist that lingered in the clammy air. I remembered that Erika had told me simply to speak the words until I felt something, and then to record it on my phone. Astonishingly, this worked – it was almost as if everything that was happening to the glacier was also happening to me. Yet again my emotions welled over, and tears followed.

On stage at the Hay Festival, my fears evaporated as I became the glacier. To be able to express long-suppressed emotions – to be freed from the straitjacket of scientific convention – felt incredibly cathartic, and I was stunned when audience members approached me afterwards and revealed that they'd wept as the glacier sickened, as it poisoned its river, and as it finally died. They wanted to know, what could they do?

One thing I've learnt is that as human beings, we are inseparable from our glaciers. Every individual will be affected by glacier shrinkage or loss in coming years, from the farming communities of the Peruvian Andes, to the halibut fishermen off Greenland's western shores, or the inhabitants of low-lying islands in the Pacific. The Earth has cycled through many extreme phases of climate, but what we are witnessing now is unprecedented in Earth's history, not just human history, and it has largely happened during the last century. Whatever your view of the role of fossil fuels in maintaining economic prosperity, the greatest loser in this game will be humankind.

I'm an optimist at heart, though, and I believe that as human beings we have an enormous capacity to change how we live. Take 2020, for instance, when people willingly went into lockdown to safeguard the lives of millions at risk from Covid-19. The risk from melting glaciers certainly rivals Covid-19 – during the twenty-first century more than a billion people will probably be affected by the repercussions of glacier melting, including rising sea levels[17] and declining water supply in major river basins.[18] Maybe our situation with glaciers is no different to a global pandemic, if we choose to see it that way.

# Afterword

## A Fork in the Path

Every glacier that I've written about has shrunk during the twenty-five years that I've been studying them, to different degrees depending on how much the air and oceans around them are warming, and the character and circumstances of each individual glacier. The same is true of the overwhelming majority of glaciers in parts of the world I've not had the fortune to visit. Small mountain glaciers, both in temperate and tropical climes, have perhaps fared the worst so far, as their small size means that they react quickly to changes in climate. Some of these have lost a third of their area over my lifetime alone, and will be all but vanished by the end of this century. Time has nearly run out for them.

Our ice sheets are initially slower to respond to the planet's warming; this is because they create their own climate, and any change in snowfall and melt on their surfaces can take some time to register as a change in the position of their fronts. However, where our ice sheets end in the ocean there are processes of instability which mean that once the warming has started to take hold, it can act like a runaway train – borne out by the accelerating collapse of floating ice tongues and ice shelves around the Greenland and Antarctic Ice Sheets. If this continues and we do not curb greenhouse gas emissions, the (remote) possibility of as much as two metres of sea-level rise by 2100 rears its head, with the potential to rise to over seven metres by 2200 as the West Antarctic Ice Sheet and ice around East Antarctica become unstable – staggering.[1] These changes will continue into the future, but their scale will depend, quite frankly, on whether we're collectively and individually prepared to make huge alterations to every aspect of how we live our lives – from what we eat, to how we heat our homes, to how much we travel and by what means, and so on. In essence, we stand at a fork in the path which will determine the fate of the glaciers. Is this really the last chapter for them?

In August 2019, while I was in Peru, a 'funeral' was held for Okjökull, the first Icelandic glacier to be lost to climate change – a sombre event attended by one hundred people, including the nation's prime minister – and a commemorative plaque was fixed to a boulder once entombed in its dirty basal layers. The inscription on the plaque pithily expresses the reality of this moment of unprecedented peril, when it is still *just* feasible to reduce our carbon emissions and, as a result, save some of our glaciers:

### A letter to the future

Ok is the first Icelandic glacier to lose its status as a glacier.
In the next 200 years all our glaciers are expected to follow
the same path.
This monument is to acknowledge that we know
what is happening and what needs to be done.
Only you know if we did it.
August 2019
415 ppm $CO_2$

# A Glacial Glossary

**Anthropocene** proposed period of Earth's history influenced by human activity. There is debate over when it started, ranging from several thousand years ago to the Industrial Revolution in the late nineteenth century or even to 1950 with the first nuclear bomb tests. It lasts up until the present day.

**Arête** knife-edged ridge created by the erosion of glaciers on opposite sides of a mountain, leaving behind only a long, narrow fin of rock.

**Autotroph** (could be a **phototroph** or **chemotroph**) from the ancient Greek, meaning 'self-nourishing'. A form of life which is able to produce its own food. It may do this using solar energy ('phototroph') or chemical energy ('chemotroph').

**Bacteria** microscopic, single-celled organisms that thrive in many different types of environments on Earth.

**Basal sliding** mechanism of glacier flow by which the base of the glacier slides over its bedrock, lubricated by water. Only occurs in temperate or polythermal glaciers.

**Carbon dioxide** a molecule comprising one carbon atom and two oxygen atoms, found as a gas at room temperature and pressure. It is present in low (but rising) concentrations in the Earth's atmosphere and is a greenhouse gas.

**Cenozoic** geological era spanning roughly the last sixty-five million years.

**Channelized drainage system** network of interconnected channels between the glacier bed and its sole, which conveys meltwater rapidly to the glacier snout.

**Cirque** small glacier that sits within glacially eroded armchair depressions high up in the mountains. Cirque glaciers can often be healthier than valley glaciers since they receive extra snow from the side walls of their rocky hollows.

**Cold-based glacier** glaciers (often small and thin) which are frozen to their beds. Usually found in the polar regions.

**Concentration** the amount of a substance (e.g. a gas, or a chemical) present in a known amount of medium (e.g. water, air). For example, a gas has a concentration in the medium of air, which might be expressed as the number of molecules of the specific gas present in a total number of molecules of gases in air (often as 'parts per million'). A chemical might have a concentration dissolved in a volume of water, expressed in units of 'grams per litre'.

**Conduit** ice-walled channel often found within a glacier or beneath it, which transports meltwater rapidly and efficiently. Dominant in peak summer.

**Cryoconite hole** cylindrical ice-walled hole with a thin layer of dark sediment on its bottom (called 'cryoconite'). It is formed because the dark sediment in the ice absorbs more radiation from the sun than the ice around it, and heats up, melting its way into the ice.

**Debris-covered glacier** glacier which is covered, to varying degrees, in debris in its lower trunk. Commonly found in the Andes and in high-mountain Asia. Debris-covered glaciers may slowly transition to 'rock glaciers' over time, where the latter are completely covered in rock, but ice is present in between the rock material, allowing the glacier to flow.

**Distributed system** slow-flowing, inefficient pathways that transport meltwater at the bed of a glacier, often zones bordering conduits/channels. Dominant in winter/spring but retained in the upper parts of glaciers that are snow-covered or have only small amounts of surface melting.

**Eemian** the very last interglacial (warm) period during the Quaternary at around 120,000 years ago. Showing warmer air temperatures than the present interglacial temperatures, than the present interglacial

(Holocene) and so a good analogue for where we might end up in the future. Sea level was around six metres higher, which is a bit of a worry.

**Electrical conductivity** refers to the capacity of a material to carry an electrical current. For a body of water this partly depends on how many charged particles (ions) are floating around which can carry the current.

**Englacial zone** the zone between the glacier and its bed – largely a solid mass of ice.

**Glacial period** period during the Quaternary when climate was cold and ice sheets were present across northern Europe, America, Greenland and Antarctica.

**Glacier forefield** area in front of the glacier, usually unvegetated and strewn with glacially transported rock material of different sizes, through which the proglacial river flows.

**GLOF** stands for Glacier Lake Outburst Flood. Caused when a lake pinned at the glacier margin, or behind one of its moraines, suddenly bursts its banks and causes freak flooding.

**Heterotroph** from the ancient Greek, meaning 'other-nourishing'. A form of life which cannot produce its own food and relies on consuming food produced by other organisms. Humans are heterotrophs.

**Holocene** our most recent interglacial, which started just over 10,000 years ago. Some say we are still in it, others say we have created our own epoch called the Anthropocene.

**Hot-water drilling** clean and popular technique used to bore a hole between a glacier surface and its bed, using high-pressure hot water and steam.

**Hydrate** (or '**clathrate hydrate**') a type of solid that forms under cold, high-pressure conditions, frozen water molecules form a cage-like structure around a 'guest' gas molecule. A good example is methane hydrate. You can imagine that as you warm the hydrate up, the water molecules melt and the methane (or other gas) is released from its cage.

**Ice age** cooler periods in Earth's history when glaciers and ice sheets expanded. People recognize five ice ages over Earth's four and a half billion year history, with the latest starting two million years ago (the Quaternary) and which we are still hovering in (just . . . ).

**Ice deformation** a process of ice flow that happens in all glaciers, caused by the slow deformation and dislocation of ice crystals under pressure.

**Ice margin** word used to describe the very edge or front of the glacier.

**Ice sheet** mass of glacier ice that largely blankets the underlying mountains and valleys (generally more than 50,000 km² in area).

**Icefall** highly crevassed zone of a glacier where ice flows rapidly down a very steep rockface. A bit like a waterfall but of ice.

**Inorganic carbon** carbon that is not bound up with living things and instead is found in things like rocks and minerals. For example, carbon found in carbon dioxide is a form of inorganic carbon, but becomes organic carbon when taken up by plants via photosynthesis and forms parts of the plant cells.

**Interglacial period** warm period of some tens of thousands of years during the Quaternary. Interspersed with glacial periods.

**Ion** an atom or a molecule that has an electrical charge, formed when it gives up or accepts a negatively charged particle called an electron.

**Isotope** different forms of a single element (e.g. oxygen) that have a different mass, owing to the fact that they have a different number of neutrons (the non-charged particles that hang out in the nucleus of an element). Oxygen, for example, has three isotopes which have masses of 16, 17 and 18. They are all still oxygen.

**Katabatic wind** a type of cold wind that flows downslope in the mountains due to gravity.

**Lake-terminating glacier** glacier whose tongue ends in a lake.

**Methane** a molecule comprising one carbon atom and four hydrogen atoms. It exists as a gas at room temperature and pressure, but can be present in solid form ('hydrate') under cold, high-pressure conditions. It is a greenhouse gas with around twenty to thirty times greater warming strength than carbon dioxide (over 100 years).

**Microbe** a very tiny form of life, only visible under a microscope, and often no more than a single cell, probably the first type of life form to have evolved on Earth.

**Molecule** group of two or more atoms happily bonded together.

**Moraine** accumulation of glacial debris (sand, stones, boulders) released from the glacier when present in one place and left behind when the glacier retreats. Can be medial (in the middle of the glacier), terminal (at the front of the glacier) or lateral (at the sides of the glacier).

**Moulin** hole in the ice, creating a vertical shaft (like a sink hole in limestone/karst landscapes) by which meltwater flows to the glacier bed. Usually formed when crevasses become engorged as meltwater flows down them.

**Naled** mass of ice present in glacier forefields, formed by the freezing of successive fractions of meltwater as it continues to flow over winter in polar climates (also called icing, or aufeis).

**Nutrient** essential substance required by life to thrive.

**Organic carbon** carbon that is bound up with living things, like plants and animals.

**Organic matter** carbon-based compounds formed from the organic remains of living things which are in the process of decaying and decomposing.

**Permafrost** ground that has a temperature below zero degrees Celsius for a year or more (usually the ground is also frozen).

**pH** a measure of how acidic or alkaline a water is. Acidic solutions have a lower pH. The pH scale runs from 0 to 14, where 7 is neutral, anything less than 7 is acidic and more than 7 is alkaline.

**Phytoplankton** tiny plant-like organisms found floating in salty or freshwater environments, such as the ocean or lakes, which can only be seen under the microscope. They use the sun to make their own food via photosynthesis.

**Polythermal** (or **polythermal-based**) **glacier** glacier whose ice is a mix of temperatures all year around. Usually these glaciers are in the polar regions and have a sub-freezing surface layer and margins and a warm, temperate core. A bit like a jammy donut.

**Pressure melting** the process by which ice melts due to the application of pressure, which shifts its melting point to a little below zero degrees Celsius.

**Proglacial zone** zone located immediately in front of the glacier (also '**glacier forefield**').

**Quaternary** geological time slice that covers roughly the last two million years and is characterized by regular cold and warm periods (glacial-interglacial cycles).

**Serac** sharp pinnacle of ice found where ice has flowed fast downslope and fractured to create many intersecting crevasses (e.g. in icefalls).

**Snout** the very downslope end of a glacier (i.e. the glacier's nose).

**Subglacial zone** the part of the glacier between the glacier's ice sole and the rock beneath it.

**Sulphide minerals** minerals found in rocks made up of sulphur and metal (often iron). Pyrite ($FeS_2$) is the most common and is released from rocks by glacial erosion.

**Supraglacial zone** zone of the glacier that covers the entire glacier surface.

**Temperate** (or **warm-based glacier**) glacier whose ice is at the melting point and able to support liquid water throughout. May develop a temporary surface layer of sub-freezing ice in winter.

**Tidewater glacier** (or **marine-terminating glacier**) glacier whose tongue terminates in the ocean.

**Tongue** lower trunk of a glacier.

**Valley glacier** glacier that flows downslope via a well-defined valley.

**Weathering** a collection of processes (physical and chemical) which result in the breakdown of rock over time.

# Notes

## Introduction

1 IPCC (2018). *Global Warming of 1.5 °C: An IPCC Special Report on the impacts of global warming of 1.5 °C above pre-industrial levels and related global greenhouse gas emission pathways, in the context of strengthening the global response to the threat of climate change, sustainable development and efforts to eradicate poverty.*

2 UNEP (2019). *Emissions Gap Report 2019*. Nairobi, UNEP.

## 1. Glimpses of an Underworld

1 Carozzi, A. V. (1966). 'Agassiz's amazing geological speculation: the Ice-Age', *Studies in Romanticism* 5 (2): 57–83.

2 Imbrie, J. and K. P. Imbrie (1986). *Ice Ages: Solving the Mystery*. Cambridge, MA: Harvard University Press.

3 Hubbard, A., et al. (2000). 'Glacier mass-balance determination by remote sensing and high-resolution modelling', *Journal of Glaciology* 46 (154): 491–8.

4 Mair, D., et al. (2002). 'Influence of subglacial drainage system evolution on glacier surface motion: Haut Glacier d'Arolla, Switzerland', *Journal of Geophysical Research: Solid Earth* 107 (B8): Doi:10.1029/2001JB000514.

5 Campbell, R. B. (2007). *In Darkest Alaska: Travels and Empire along the Inside Passage*. Philadelphia, PA: University of Philadelphia Press.

6  Hubbard, B. P., et al. (1995). 'Borehole water-level variations and the structure of the subglacial hydrological system of Haut Glacier d'Arolla, Valais, Switzerland', *Journal of Glaciology* 41 (139): 572–83.

7  Nienow, P., et al. (1998). 'Seasonal changes in the morphology of the subglacial drainage system, Haut Glacier d'Arolla, Switzerland', *Earth Surface Processes and Landforms* 23 (9): 825–43.

8  Hubbard et al. (1995).

9  Iken, A., et al. (1983). 'The uplift of Unteraargletscher at the beginning of the melt season – a consequence of water storage at the bed?', *Journal of Glaciology* 29 (101): 28–47.

10  Von Hardenberg, A., et al. (2004). 'Horn growth but not asymmetry heralds the onset of senescence in male Alpine ibex (*Capra ibex*)', *Journal of Zoology* 263: 425–32.

11  http://oldeuropeanculture.blogspot.com/2016/12/goat.html.

12  Maixner, F., et al. (2018). 'The Iceman's last meal consisted of fat, wild meat, and cereals', *Current Biology* 28 (14): 2,348–55.

13  Sharp, M., et al. (1999). 'Widespread bacterial populations at glacier beds and their relationship to rock weathering and carbon cycling', *Geology* 27 (2): 107–10.

14  Fischer, M., et al. (2015). 'Surface elevation and mass changes of all Swiss glaciers 1980–2010', *The Cryosphere* 9 (2): 525–40.

15  Berner, R. A. (2003). 'The long-term carbon cycle, fossil fuels and atmospheric composition', *Nature* (426): 323–6.

16  Zachos, J. C., et al. (2008). 'An early Cenozoic perspective on greenhouse warming and carbon-cycle dynamics', *Nature* 451 (7176): 279–83.

17  Guardian, T. (2012). 'Greenhouse gas levels pass symbolic 400ppm $CO_2$ milestone', https://www.esrl.noaa.gov/gmd/ccgg/trends/weekly.html

18  Raymo, M. E., et al. (1988). 'Influence of late Cenozoic mountain building on ocean geochemical cycles', *Geology* 16 (7): 649–53.

19 Pearson, P. N., et al. (2009). 'Atmospheric carbon dioxide through the Eocene-Oligocene climate transition', *Nature* 461 (7267): 1,110–13.

20 Hays, J. D., et al. (1976). 'Variations in the Earth's orbit: pacemaker of the Ice Ages', *Science* 194 (4270): 1,121–32.

21 Maslin, M. A., et al. (1998). 'The contribution of orbital forcing to the progressive intensification of northern hemisphere glaciation', *Quaternary Science Reviews* 17 (4): 411–26.

22 Grove, J. (1988). *The Little Ice Age*. Ann Arbor, MI: University of Michigan/London: Methuen.

23 EPICA Community Members (2004). 'Eight glacial cycles from an Antarctic ice core', *Nature* 429: 623–8.

24 Miller, K. G., et al. (2012). 'High tide of the warm Pliocene: implications of global sea level for Antarctic deglaciation', *Geology* 40 (5): 407–10.

25 Haywood, A. M., et al. (2013). 'Large-scale features of Pliocene climate: results from the Pliocene Model Intercomparison Project', *Climate of the Past* 9 (1): 191–209.

26 Foster, G. L., et al. (2017). 'Future climate forcing potentially without precedent in the last 420 million years', *Nature Communications* 8: Doi:10.1038/ncomms14845.

27 Radić, V., et al. (2014). 'Regional and global projections of twenty-first-century glacier mass changes in response to climate scenarios from global climate models', *Climate Dynamics* 42 (1): 37–58; Zekollari, H., et al. (2019). 'Modelling the future evolution of glaciers in the European Alps under the EURO-CORDEX RCM ensemble', *The Cryosphere* 13 (4): 1,125–46.

## 2. Bears, Bears Everywhere

1 Derocher, A. (2012). *Polar Bears: A Complete Guide to their Biology and Behaviour*. Baltimore, MD: Johns Hopkins University Press.

2 Meredith, M., et al. (2019). Chapter 3, 'Polar Regions', in *IPCC Special Report on the Ocean and Cryosphere in a Changing Climate*, ed. H. O. Pörtner, D. C. Roberts, V. Masson-Delmotte et al.: 118.

3 Liestøl, O. (1993). 'Glaciers of Svalbard, Norway: Satellite Image Atlas of Glaciers of the World', in *Glaciers of Europe*, ed. R. S. Williams and J. G. Ferrigno, US Geological Survey Professional Paper 1386–E: 127–52.

4 Åkerman, J. (1982). *Studies on Naledi (Icings) in West Spitsbergen*. Proceedings of the 4th Canadian Permafrost Conference, Ottawa, National Research Council of Canada: 189–202.

5 Liestøl, O. (1969). 'Glacier surges in West Spitsbergen', *Canadian Journal of Earth Sciences* 6 (4): 895–7; Baranowski, S. (1983). 'Naled ice in front of some Spitsbergen glaciers', *Journal of Glaciology* 28 (98): 211–14.

6 Sorensen, A. C., et al. (2018). 'Neandertal fire-making technology inferred from microwear analysis', *Scientific Reports* 8 (1): Doi:10.1038/s41598-018-28342-99.

7 Skidmore, M. and M. Sharp (1995). 'Drainage system behaviour of a High-Arctic polythermal glacier', *Annals of Glaciology* 28: 209–15.

8 Wadham, J. L., et al. (2001). 'Evidence for seasonal subglacial outburst events at a polythermal glacier, Finsterwalderbreen, Svalbard', *Hydrological Processes* 15 (12): 2,259–80.

9 Nuttall, A.-M. and R. Hodgkins (2005). 'Temporal variations in flow velocity at Finsterwalderbreen, a Svalbard surge-type glacier', *Annals of Glaciology* 42: 71–6.

10 Prop, J., et al. (2015). 'Climate change and the increasing impact of polar bears on bird populations', *Frontiers in Ecology and Evolution* 25: Doi:10.3389/fevo.2015.00033.

11 Bottrell, S. H. and M. Tranter (2002). 'Sulphide oxidation under partially anoxic conditions at the bed of the Haut Glacier d'Arolla, Switzerland', *Hydrological Processes* 16 (12): 2,363–8.

12  Wadham, J. L., et al. (2004). 'Stable isotope evidence for microbial sulphate reduction at the bed of a polythermal high Arctic glacier', *Earth and Planetary Science Letters* 219 (3–4): 341–55.

13  Meredith et al. (2019).

14  Hanssen-Bauer, I., et al. (2018). 'Climate in Svalbard 2100 – a knowledge base for climate adaptation', Norwegian Environment Agency.

15  Haug, T., et al. (2017). 'Future harvest of living resources in the Arctic Ocean north of the Nordic and Barents Seas: a review of possibilities and constraints', *Fisheries Research* 188: 38–57.

16  Onarheim, I. H., et al. (2014). 'Loss of sea ice during winter north of Svalbard', *Tellus A: Dynamic Meteorology and Oceanography* 66 (1): Doi:10.3402/tellusa.v66.23933.

17  Muckenhuber, S., et al. (2016). 'Sea ice cover in Isfjorden and Hornsund, Svalbard (2000–2014) from remote sensing data', *The Cryosphere* 10 (1): 149–58.

## *3. Plumbing the Depths*

1  Nuttall, M. (2010). 'Anticipation, climate change, and movement in Greenland', *Études/Inuit/Studies* 34 (1): 21–37.

2  Kane, N. (2019). *History of the Vikings and Norse Culture.* Spangenhelm Publishing.

3  Nedkvitne, A. (2019). *Norse Greenland: Viking Peasants in the Arctic.* Oxford: Routledge.

4  Ibid.

5  Wells, N. C. (2016). 'The North Atlantic Ocean and climate change in the UK and northern Europe', *Weather* 71 (1): 3–6.

6  Rea, B. R., et al. (2018). 'Extensive marine-terminating ice sheets in Europe from 2.5 million years ago', *Science Advances* 4 (6): Doi:10.1126/sciadv.aar8327.

7  Lambeck, K., et al. (2014). 'Sea level and global ice volumes from the Last Glacial Maximum to the Holocene', *Proceedings of the National Academy of Sciences* 111 (43): 15,296–303.

8  Oppenheimer, M., et al. (2019). Chapter 4: 'Sea Level Rise and Implications for Low-lying Islands, Coasts and Communities', in *IPCC Special Report on the Ocean and Cryosphere in a Changing Climate*, ed. H.-O. Pörtner, D. C. Roberts, V. Masson-Delmotte et al.: 321–445.

9  Shennan, I., et al. (2006). 'Relative sea-level changes, glacial isostatic modelling and ice-sheet reconstructions from the British Isles since the Last Glacial Maximum', *Journal of Quaternary Science* 21 (6): 585–99.

10  Meredith et al. (2019).

11  Tedesco, M. and X. Fettweis (2020). 'Unprecedented atmospheric conditions (1948–2019) drive the 2019 exceptional melting season over the Greenland ice sheet', *The Cryosphere* 14 (4): 1,209–23.

12  Rignot, E., et al. (2012). 'Spreading of warm ocean waters around Greenland as a possible cause for glacier acceleration', *Annals of Glaciology* 53 (60): 257–66.

13  Howat, I. M. and A. Eddy (2011). 'Multi-decadal retreat of Greenland's marine-terminating glaciers', *Journal of Glaciology* 57 (203): 389–96.

14  Ibid.

15  Oppenheimer et al. (2019).

16  Ibid.

17  Duncombe, J. (2019). 'Greenland ice sheet beats all-time 1-day melt record', *EOS, Transactions of the American Geophysical Union* 100: Doi.org/10.1029/2019EO130349.

18  Meredith et al. (2019).

19  Stibal, M., et al. (2017). 'Algae drive enhanced darkening of bare ice on the Greenland Ice Sheet', *Geophysical Research Letters* 44 (22): 11,463–71.

20  Williamson, C. J., et al. (2019). 'Glacier algae: a dark past and a darker future', *Frontiers in Microbiology* 10 (524): 10.3389/fmicb.2019.00524.

21 Chandler, D. et al. (2013). 'Evolution of the subglacial drainage system beneath the Greenland Ice Sheet revealed by tracers', *Nature Geoscience* 6 (3): 195–8.

22 Ibid.

23 Tedstone, A. J., et al. (2015). 'Decadal slowdown of a land-terminating sector of the Greenland Ice Sheet despite warming', *Nature* 526 (7575): 692–5.

24 Davison, B. J., et al. (2019). 'The influence of hydrology on the dynamics of land-terminating sectors of the Greenland Ice Sheet', *Frontiers in Earth Science* 7 (10): Doi:10.3389/feart.2019.00010.

25 Ibid.

26 Cowton, T., et al. (2012). 'Rapid erosion beneath the Greenland ice sheet', *Geology* 40 (4): 343–6.

27 Hudson, B., et al. (2014). 'MODIS observed increase in duration and spatial extent of sediment plumes in Greenland fjords', *The Cryosphere* 8 (4): 1,161–76.

28 Meire, L., et al. (2017). 'Marine-terminating glaciers sustain high productivity in Greenland fjords', *Global Change Biology* 23: 5,344–57; Middelbo, A. B., et al. (2018). 'Impact of glacial meltwater on spatiotemporal distribution of copepods and their grazing impact in Young Sound NE, Greenland', *Limnology and Oceanography* 63 (1): 322–36.

29 Cowton, T. R., et al. (2018). 'Linear response of east Greenland's tidewater glaciers to ocean/atmosphere warming', *Proceedings of the National Academy of Sciences* 115 (31): 7,907–12.

30 Juul-Pedersen, T., et al. (2015). 'Seasonal and interannual phytoplankton production in a sub-Arctic tidewater outlet glacier fjord, SW Greenland', *Marine Ecology Progress Series* 524: 27–38.

31 Meire, L., et al. (2016). 'Spring bloom dynamics in a subarctic fjord influenced by tidewater outlet glaciers (Godthåbsfjord, SW Greenland)', *Journal of Geophysical Research-Biogeosciences* 121: 1, 581–92.

32 Meire et al. (2017); Juul-Pedersen et al. (2015).

33 ICES (2015). *Report of the North-Western Working Group (NWWG)*, Copenhagen.

34 Meire et al. (2017).

35 Hendry, K. R., et al. (2019). 'The biogeochemical impact of glacial meltwater from Southwest Greenland', *Progress in Oceanography* 176: Doi: 10.1016/j.pocean.2019.102126.

36 Hawkings, J., et al. (2014). 'Ice sheets as a significant source of highly reactive nanoparticulate iron to the oceans', *Nature Communications* 5: Doi:10.1038/ncomms4929; Hawkings, J., et al. (2017). 'Ice sheets as a missing source of silica to the world's oceans', ibid.: 8: Doi:10.1038/ncomms14198; Hawkings, J., et al. (2016). 'The Greenland Ice Sheet as a hot spot of phosphorus weathering and export in the Arctic', *Global Biogeochemical Cycles* 30 (2): 191–210.

37 Duprat, L. P. A. M., et al. (2016). 'Enhanced Southern Ocean marine productivity due to fertilization by giant icebergs', *Nature Geoscience* 9 (3): 219–21.

38 Howat and Eddy (2011).

39 Meredith et al. (2019).

40 Sonne, B. (2017). *Worldviews of the Greenlanders: An Inuit Arctic Perspective*. Fairbanks, AK: University of Alaska Press.

41 ACIA (2005). *Arctic Climate Impact Assessment (ACIA)*. Cambridge: Cambridge University Press.

42 Nuttall (2010).

43 Hastrup, K. (2018). 'A history of climate change: Inughuit responses to changing ice conditions in North-West Greenland', *Climatic Change* 151: 67–78.

44 Ross, J. (1819). *Voyage of Discovery, made under the orders of Admiralty, in his Majesty's ships Isabelle and Alexander, for the Purpose of Exploring Baffin's Bay, and inquiring into the probability of a North-West Passage*. London: C. U. Press.

45 Hastrup (2018).

46 Meredith et al. (2019).

47 US Fish & Wildlife Service, 1995. *Muskox: Ovibos Moschatus*, Biologue Series, University of Minnesota.

48 Lasher, G. E. and Y. Axford (2019). 'Medieval warmth confirmed at the Norse Eastern Settlement in Greenland', *Geology* 47 (3): 267–70.

49 McGovern, T. H. (1991). 'Climate, correlation, and causation in Norse Greenland', *Arctic Anthropology* 28 (2): 77–100.

50 Star, B., et al. (2018). 'Ancient DNA reveals the chronology of walrus ivory trade from Norse Greenland', *Proceedings of the Royal Society B: Biological Sciences* 285 (1884): Doi:10.1098/rspb. 2018.0978; Barrett, J. H., et al. (2020). 'Ecological globalisation, serial depletion and the medieval trade of walrus rostra', *Quaternary Science Reviews* 229: Doi.org/10.1016/j.quascirev.2019.106122.

51 Barrett et al. (2020); Star et al. (2018).

52 McGovern, T. H. (2018). 'Greenland's lost Norse: parables of adaptation from the North Atlantic', in *Polar Geopolitics: A Podcast on the Arctic and Antarctica*. E. Bagley. http://www.podbean.com/eu/pb-pixwv-a83cdb.

53 Gulløv, H. C. (2008). 'The nature of contact between native Greenlanders and Norse', *Journal of the North Atlantic* 1: 16–24.

54 Barrett et al. (2020).

55 Kintisch, E. (2016). 'Why did Greenland's Vikings disappear?', *Science: Archaeology and Human Evolution*, Doi:10.1126/science. aal0363.

56 Dugmore, A. J., et al. (2012). 'Cultural adaptation, compounding vulnerabilities and conjunctures in Norse Greenland', *Proceedings of the National Academy of Sciences* 109 (10): 3,658–63.

57 McGovern (2018).

## 4. *Life at the Extremes*

1 Ainley, D. G. (2002). *The Adélie Penguin: Bellwether of Climate Change*, New York: Columbia University Press.

2 Ibid.

3 Fountain, A. G., et al. (2016). 'Glaciers in equilibrium, McMurdo Dry Valleys, Antarctica', *Journal of Glaciology* 62 (235): 976–89.

4 Ibid.

5 Koch, P. L., et al. (2019). 'Mummified and skeletal southern elephant seals (*Mirounga leonina*) from the Victoria Land Coast, Ross Sea, Antarctica', *Marine Mammal Science* 35 (3): 934–56.

6 Scott, R. F. (1905). *The Voyage of the Discovery VII*. London: Macmillan; Priscu, J. C. (1999). 'Life in the Valley of the "Dead"', *BioScience* 49 (12): 959.

7 Fountain, A. G., et al. (2004). 'Evolution of cryoconite holes and their contribution to meltwater runoff from glaciers in the McMurdo Dry Valleys, Antarctica', *Journal of Glaciology* 50 (168): 35–45.

8 Tranter, M., et al. (2010). 'The biogeochemistry and hydrology of McMurdo Dry Valley glaciers: is there life on martian ice now?': 195–220, in *Life in Antarctic Deserts and Other Cold Dry Environments: Astrobiological Analogs*, ed. P. T. Doran, W. B. Lyons and D. M. McKnight. Cambridge: Cambridge University Press.

9 Ibid.; Priscu (1999).

10 Bagshaw, E. A., et al. (2016). 'Response of Antarctic cryoconite microbial communities to light', *FEMS Microbiology Ecology* 92 (6): Doi.org/10.1093/femsec/fiw076.

11 Ibid.

12 Dubnick, A., et al. (2017). 'Trickle or treat: the dynamics of nutrient export from polar glaciers', *Hydrological Processes* 31 (9): 1, 776–89.

13  Gooseff, M. N., et al. (2017). 'Decadal ecosystem response to an anomalous melt season in a polar desert in Antarctica', *Nature Ecology & Evolution* 1 (9): 1,334–8.

14  Dubnick et al. (2017); Bagshaw, E. A., et al. (2013). 'Do cryoconite holes have the potential to be significant sources of C, N, and P to downstream depauperate ecosystems of Taylor Valley, Antarctica?', *Arctic, Antarctic, and Alpine Research* 45 (4): 440–54.

15  Fritsen, C. H. and J. C. Priscu (1999). 'Seasonal change in the optical properties of the permanent ice cover on Lake Bonney, Antarctica: consequences for lake productivity and phytoplankton dynamics', *Limnology and Oceanography* 44: 447–54.

16  Mikucki, J. A., et al. (2009). 'A contemporary microbially maintained subglacial ferrous "ocean"', *Science* 324 (5925): 397–400.

17  Naylor, S., et al. (2008). 'The IGY and the ice sheet: surveying Antarctica', *Journal of Historical Geography* 34 (4): 574–95.

18  Siegert, M. J. (2018). 'A 60-year international history of Antarctic subglacial lake exploration', *Geological Society, London, Special Publications* 461 (1): 7–21.

19  Pattyn, F. (2010). 'Antarctic subglacial conditions inferred from a hybrid ice sheet/ice stream model', *Earth and Planetary Science Letters* 295 (3–4): 451–61.

20  Priscu, J., et al. (2008). 'Antarctic subglacial water: origin, evolution and ecology', in *Polar Lakes and Rivers*, ed. W. F. Vincent and J. Laybourne-Parry, New York: Oxford University Press: 119–36; Siegert, M. J., et al. (2016). 'Recent advances in understanding Antarctic subglacial lakes and hydrology', *Philosophical Transactions of the Royal Society A* 374 (2059): Doi:10.1098/rsta.2014.0306.

21  Escutia, C., et al. (2019). 'Keeping an eye on Antarctic Ice Sheet stability', *Oceanography* 32 (1): 32–46.

22  Krasnopolsky, V. A., et al. (2004). 'Detection of methane in the martian atmosphere: evidence for life?', *Icarus* 172 (2): 537–47.

23  Stibal, M., et al. (2012). 'Methanogenic potential of Arctic and Antarctic subglacial environments with contrasting organic carbon sources', *Global Change Biology* 18 (11): 3,332–45.

24  Wadham, J. L., et al. (2012). 'Potential methane reservoirs beneath Antarctica', *Nature* 488 (7413): 633–7.

25  Michaud, A. B., et al. (2017). 'Microbial oxidation as a methane sink beneath the West Antarctic Ice Sheet', *Nature Geoscience* 10 (8): 582–6.

26  Wadham, J. L., et al. (2013). 'The potential role of the Antarctic Ice Sheet in global biogeochemical cycles', *Earth and Environmental Science Transactions of the Royal Society of Edinburgh* 104 (1): 55–67.

27  Maule, C. F., et al. (2005). 'Heat flux anomalies in Antarctica revealed by satellite magnetic data', *Science* 309 (5733): 464–7.

28  van Wyk de Vries, M., et al. (2018). 'A new volcanic province: an inventory of subglacial volcanoes in West Antarctica', *Geological Society, London, Special Publications* 461 (1): 231–48.

29  Wadham et al. (2012).

30  Pritchard, H. D., et al. (2012). 'Antarctic ice-sheet loss driven by basal melting of ice shelves', *Nature* 484 (7395): 502–5.

31  Schmidtko, S., et al. (2014). 'Multidecadal warming of Antarctic waters', *Science* 346 (6214): 1,227–31.

32  Thompson, D. W. J. and S. Solomon (2002). 'Interpretation of recent southern hemisphere climate change', *Science* 296 (5569): 895–9.

33  Lee, S. and S. B. Feldstein (2013). 'Detecting ozone- and greenhouse gas-driven wind trends with observational data', *Science* 339 (6119): Doi:10.1126/science.1225154.

34  Ibid.; Meredith et al. (2019).

35  Holland, P. R., et al. (2019). 'West Antarctic ice loss influenced by internal climate variability and anthropogenic forcing', *Nature Geoscience* 12 (9): 718–24.

36  Meredith et al. (2019).

37 Wingham, D. J., Wallis, D. W., and Shepherd, A. (2009). 'Spatial and temporal evolution of Pine Island Glacier thinning, 1995–2006', *Geophysical Research Letters* 36 (17): Doi:10.1029/2009GL039126.

38 Rignot, E., et al. (2019). 'Four decades of Antarctic Ice Sheet mass balance from 1979–2017', *Proceedings of the National Academy of Sciences* 116 (4): 1,095–103.

39 DeConto, R. M. and D. Pollard (2016). 'Contribution of Antarctica to past and future sea-level rise', *Nature* 531 (7596): 591–7; Turney, C. S. M., et al. (2020). 'Early last interglacial ocean warming drove substantial ice mass loss from Antarctica', *Proceedings of the National Academy of Sciences* 117 (8): 3,996–4,066.

40 Oppenheimer et al. (2019).

41 Andreassen, K., et al. (2017). 'Massive blow-out craters formed by hydrate-controlled methane expulsion from the Arctic seafloor', *Science* 356 (6341): 948–53.

42 'Paris Agreement', United Nations Treaty Collection, 8 July 2016.

43 Portnov, A., et al. (2016). 'Ice-sheet-driven methane storage and release in the Arctic', *Nature Communications* 7: Doi:10.1038/ncomms10314.

44 Michaud et al. (2017).

45 Lamarche-Gagnon, G., et al. (2019). 'Greenland melt drives continuous export of methane from the ice-sheet bed', *Nature* 565 (7737): 73–7.

46 Thurber, A. R., et al. (2020). 'Riddles in the cold: Antarctic endemism and microbial succession impact methane cycling in the Southern Ocean', *Proceedings of the Royal Society B: Biological Sciences* 287 (1931): Doi.org/10.1098/rspb.2020.1134.

## 5. Beware of the GLOF!

1 Millan, R., et al. (2019). 'Ice thickness and bed elevation of the northern and southern Patagonian icefields', *Geophysical Research Letters* 46 (12): 6,626–35.

2 Garreaud, R., et al. (2013). 'Large-scale control on the Patagonian climate', *Journal of Climate* 26 (1): 215–30.

3 Bendle, J. M., et al. (2019). 'Phased Patagonian Ice Sheet response to Southern Hemisphere atmospheric and oceanic warming between 18 and 17 ka', *Scientific Reports* 9 (1): Doi:10.1038/s41598-1019-39750-w.

4 Lenaerts, J. T. M., et al. (2014). 'Extreme precipitation and climate gradients in Patagonia revealed by high-resolution regional atmospheric climate modeling', *Journal of Climate* 27 (12): 4,607–21.

5 Mouginot, J. and E. Rignot (2015). 'Ice motion of the Patagonian icefields of South America: 1984–2014', *Geophysical Research Letters* 42 (5): 1,441–9.

6 Zemp, M., et al. (2019). 'Global glacier mass changes and their contributions to sea-level rise from 1961 to 2016', *Nature* 568 (7752): 382–6.

7 Rivera, A., et al. (2012). 'Little Ice Age advance and retreat of Glaciar Jorge Montt, Chilean Patagonia', *Climate of the Past* 8 (2): 403–14.

8 Wilson, R., et al. (2018). 'Glacial lakes of the Central and Patagonian Andes', *Global and Planetary Change* 162: 275–91.

9 Carrivick, J. L. and D. J. Quincey (2014). 'Progressive increase in number and volume of ice-marginal lakes on the western margin of the Greenland Ice Sheet', *Global and Planetary Change* 116: 156–63.

10 Maharjan, S. B., et al. (2018). *The Status of Glacial Lakes in the Hindu Kush Himalaya*. Kathmandu: ICIMOD.

11 Moss, C. (2016). *Patagonia: A Cultural History* (Landscapes of the Imagination), London: Andrews.

12 Neruda, P. (2000). *Canto General*, 50th Anniversary Edition, trans. Jack Schmitt. Berkeley, CA: University of California Press, p. 227.

13 Palmer, J. (2019). 'The dangers of glacial lake floods: pioneering and capitulation', *EOS, Transactions of the American Geophysical Union* 100: Doi:org/10.1029/2019EO116807.

14 Harrison, S., et al. (2018). 'Climate change and the global pattern of moraine-dammed glacial lake outburst floods', *The Cryosphere* 12 (4): 1,195–209.

15 Davies, B. J. and N. F. Glasser (2012). 'Accelerating shrinkage of Patagonian glaciers from the Little Ice Age (~AD 1870) to 2011', *Journal of Glaciology* 58 (212): 1,063–84.

16 Pryer, H., et al. (2020). 'Impact of glacial cover on riverine silicon and iron export to downstream ecosystems', *Global Biogeochemical Cycles* 34 (12): Doi.org/10.1029/2020SB006611.

17 Piret, L., et al. 'High-resolution fjord sediment record of a retreating glacier with growing intermediate proglacial lake (Steffen Fjord, Chile)', *Earth Surface Processes and Landforms*: Doi.org/10.1002/esp.5015.

18 Iriarte, J. L., et al. (2018). 'Low spring primary production and microplankton carbon biomass in Sub-Antarctic Patagonian channels and fjords (50–53°S)', *Arctic, Antarctic, and Alpine Research* 50 (1): Doi:10.1080/15230430.2018.1525186.

19 Cuevas, L. A., et al. (2019). 'Interplay between freshwater discharge and oceanic waters modulates phytoplankton size-structure in fjords and channel systems of the Chilean Patagonia', *Progress in Oceanography* 173: 103–13; González, H. E., et al. (2013). 'Land–ocean gradient in haline stratification and its effects on plankton dynamics and trophic carbon fluxes in Chilean Patagonian fjords (47–50°S)', *Progress in Oceanography* 119: 32–47.

20 Pryer et al. (2020).

21 Dussaillant J. A., et al. (2012). 'Hydrological regime of remote catchments with extreme gradients under accelerated change:

the Baker basin in Patagonia', *Hydrological Sciences Journal* 57 (8): 1,530–42.

22 Gillett, N. P. and D. W. J. Thompson (2003). 'Simulation of recent southern hemisphere climate change', *Science* 302 (5643): 273–5. Lee, S. and S. B. Feldstein (2013). 'Detecting ozone- and greenhouse gas-driven wind trends with observational data', *Science* 339 (6119): 563–7.

23 Lara, A., et al. (2015). 'Reconstructing streamflow variation of the Baker River from tree-rings in Northern Patagonia since 1765', *Journal of Hydrology* 529: 511–23.

## 6. *White Rivers Running Dry*

1 Wester, P., et al. (2019). *The Hindu Kush Himalaya Assessment – Mountains, Climate Change, Sustainability and People*, New York: Springer International Publishing.

2 Andermann, C., et al. (2012). 'Impact of transient groundwater storage on the discharge of Himalayan rivers', *Nature Geoscience* 5 (2): 127–32.

3 Biemans, H., et al. (2019). 'Importance of snow and glacier meltwater for agriculture on the Indo-Gangetic Plain', *Nature Sustainability* 2 (7): 594–601.

4 Wester et al. (2019).

5 Richey, A. S., et al. (2015). 'Quantifying renewable groundwater stress with GRACE', *Water Resources Research* 51 (7): 5,217–38.

6 Worldbank (1960). *The Indus Waters Treaty*.

7 Haines, D. (2017). *Rivers Divided: Indus Basin Waters in the Making of India and Pakistan*. Building the Empire, Building the Nation: Development, Legitimacy, and Hydro-Politics in Sind, 1919–1969. London: C. Hurst & Co Ltd.

8   Lutz, A. F., et al. (2014). 'Consistent increase in High Asia's runoff due to increasing glacier melt and precipitation', *Nature Climate Change* 4 (7): 587–92; Biemans et al. (2019).

9   Lutz et al. (2014).

10  Azam, M. F., et al. (2014). 'Processes governing the mass balance of Chhota Shigri Glacier (western Himalaya, India) assessed by point-scale surface energy balance measurements', *The Cryosphere* 8 (6): 2,195–217.

11  Wester et al. (2019).

12  IPCC (2013). *Climate Change 2013 – The Physical Science Basis*. Cambridge: Cambridge University Press.

13  Wester et al. (2019).

14  Fujita, K. (2008). 'Effect of precipitation seasonality on climatic sensitivity of glacier mass balance', *Earth and Planetary Science Letters* 276 (1): 14–19; Azam et al. (2014).

15  Maurer, J. M., et al. (2019). 'Acceleration of ice loss across the Himalayas over the past 40 years', *Science Advances* 5 (6); King, O., et al. (2019). 'Glacial lakes exacerbate Himalayan glacier mass loss', *Scientific Reports* 9 (1): Doi:10.1038/s41598-019-53733-x.

16  Bolch, T., et al. (2012). 'The state and fate of Himalayan glaciers', *Science* 336 (6079): 310–14.

17  Farinotti, D., et al. (2020). 'Manifestations and mechanisms of the Karakoram glacier Anomaly', *Nature Geoscience* 13 (1): 8–16.

18  Wester et al. (2019).

19  Ibid.

20  Ibid.

21  Mallet, V. (2017). *River of Life, River of Death: The Ganges and India's Future*. Oxford: Oxford University Press.

22  Tveiten, I. N. (2007). 'Glacier growing: a local response to water scarcity in Baltistan and Gilgit, Pakistan', unpublished Master's thesis, Norwegian University of Life Science.

23  Clouse, C. (2017). 'The Himalayan ice stupa: Ladakh's climate-adaptive water cache', *Journal of Architectural Education* 71 (2): 247–51.

24  Bradley, J. A., et al. (2014). 'Microbial community dynamics in the forefield of glaciers', *Proceedings of the Royal Society B: Biological Sciences* 281: Doi.org/10.1098/rspb.2014.0882.

25  Anderson, K., et al. (2020). 'Vegetation expansion in the subnival Hindu Kush Himalaya', *Global Change Biology* 26 (3): 1,608–25.

26  Wester et al. (2019).

27  Higgins, S. A., et al. (2018). 'River linking in India: downstream impacts on water discharge and suspended sediment transport to deltas', *Elementa* 6 (1): Doi.org/10.1525/elementa.269.

28  Macklin, M. G. and J. Lewin (2015). 'The rivers of civilization', *Quaternary Science Reviews* 114: 228–44.

29  Higgins et al. (2018).

## 7. The Last Ice

1  Schauwecker, S., et al. (2017). 'The freezing level in the tropical Andes, Peru: an indicator for present and future glacier extents', *Journal of Geophysical Research: Atmospheres* 122 (10): 5,172–89.

2  Schauwecker, S., et al. (2014). 'Climate trends and glacier retreat in the Cordillera Blanca, Peru, revisited', *Global and Planetary Change* 119: 85–97.

3  Seehaus, T., et al. (2019). 'Changes of the tropical glaciers throughout Peru between 2000 and 2016 – mass balance and area fluctuations', *The Cryosphere* 13 (10): 2,537–56.

4  Schauwecker et al. (2017).

5  Kaser, G., et al. (2003). 'The impact of glaciers on the runoff and the reconstruction of mass balance history from hydrological data in the tropical Cordillera Blanca, Perú', *Journal of Hydrology* 282: 130–44.

6  Milner, A. M., et al. (2017). 'Glacier shrinkage driving global changes in downstream systems', *Proceedings of the National Academy of Sciences* 114 (37): 9,770–8.

7  Margirier, A., et al. (2018). 'Role of erosion and isostasy in the Cordillera Blanca uplift: insights from landscape evolution modeling (northern Peru, Andes)', *Tectonophysics* 728–9: 119–29.

8  Gurgiser, W., et al. (2013). 'Modeling energy and mass balance of Shallap Glacier, Peru', *The Cryosphere* 7 (6): 1,787–802.

9  Petford, N. and M. P. Atherton (1992). 'Granitoid emplacement and deformation along a major crustal lineament: the Cordillera Blanca, Peru', *Tectonophysics* 205 (1): 171–85.

10 Bebbington, A. J. and J. T. Bury (2009). 'Institutional challenges for mining and sustainability in Peru', *Proceedings of the National Academy of Sciences* 106 (41): 17,296–301.

11 Durán-Alarcón, C., et al. (2015). 'Recent trends on glacier area retreat over the group of Nevados Caullaraju-Pastoruri (Cordillera Blanca, Peru) using Landsat imagery', *Journal of South American Earth Sciences* 59: 19–26.

12 Loayza-Muro, Raúl A., et al. (2013). 'Metal leaching, acidity, and altitude confine benthic macroinvertebrate community composition in Andean streams', *Environmental Toxicology and Chemistry* 33 (2): 404–11.

13 Santofimia, E., et al. (2017). 'Acid rock drainage in Nevado Pastoruri glacier area (Huascarán National Park, Perú): hydrochemical and mineralogical characterization and associated environmental implications', *Environmental Science and Pollution Research* 24 (32): 25,243–59.

14 Gurgiser et al. (2013).

15 Mark, B. G., et al. (2017). 'Glacier loss and hydro-social risks in the Peruvian Andes', *Global and Planetary Change* 159: 61–76.

16 Fraser, B. (2009). 'Climate change equals culture change in the Andes', *Scientific American*, 5 October.

17  Kulp, S. A. and B. H. Strauss (2019). 'New elevation data triple estimates of global vulnerability to sea-level rise and coastal flooding', *Nature Communications* 10 (1): Doi.org/10.1038/s41467-019-12808-z .

18  IPCC Special Report 1.5 (2018).

## *Afterword*

1  Bamber, J. L., et al. (2019). 'Ice sheet contributions to future sea-level rise from structured expert judgment', *Proceedings of the National Academy of Sciences* 116 (23): 11,195–200.

# Acknowledgements

My thanks go to the many colleagues, students and glaciers with whom I have passed endless hours organizing complex logistics, hauling gear, shivering in tents, warding off hungry polar bears and sampling icy rivers, and who have shared both joyful and dismal moments with me. The tales told in *Ice Rivers* are as much yours as mine.

I am indebted to Dr Lou Bashall (my friend and chiropractor), to Dr Darren Smyk and the many medical professionals involved in saving my brain in 2018, and to my brother Jake for his encouragement with this project and for his wizardry with words, all of whom, along with Poppy the Labrador, Pete Nienow, Anne-Marie Bremner, Till Bruckner, Patrick McGuinness, Peter Straus, and Richard Atkinson, Ania Gordon, Corina Romonti and colleagues at Penguin, formed the links in the chain that have (I hope) produced something deserving of a comfy chair, a crackling fire and a small glass of something warm and tasty.

# Picture Credits

Grateful acknowledgement is given to the following for permission to reproduce photographs: Pete Nienow, nos. 2–4 and 20; Jon Ove Hagen, no. 6; Anne-Marie Nuttall, no. 7; Grzegorz Lis, nos. 9 and 12; Megan Barnett, no. 10; Catie Butler, no. 11; Tom Kirkpatrick, no. 15; Sarah Tingey, no. 19; Miranda Thomas, no. 21; Raúl Loayza-Muro, no. 22; Rosie Bisset, nos. 24 and 25; and Jon Spaull, no. 26. Nos. 1, 5, 8, 13, 14, 16–18 and 23 were taken by the author.

ALLEN LANE
*an imprint of*
PENGUIN BOOKS

# Also Published

Joseph Sassoon, *The Global Merchants: The Enterprise and Extravagance of the Sassoon Dynasty*

Clare Chambers, *Intact: A Defence of the Unmodified Body*

Nina Power, *What Do Men Want?: Masculinity and Its Discontents*

Ivan Jablonka, *A History of Masculinity: From Patriarchy to Gender Justice*

Thomas Halliday, *Otherlands: A World in the Making*

Sofi Thanhauser, *Worn: A People's History of Clothing*

Sebastian Mallaby, *The Power Law: Venture Capital and the Art of Disruption*

David J. Chalmers, *Reality+: Virtual Worlds and the Problems of Philosophy*

Jing Tsu, *Kingdom of Characters: A Tale of Language, Obsession and Genius in Modern China*

Lewis R. Gordon, *Fear of Black Consciousness*

Leonard Mlodinow, *Emotional: The New Thinking About Feelings*

Kevin Birmingham, *The Sinner and the Saint: Dostoevsky, a Crime and Its Punishment*

Roberto Calasso, *The Book of All Books*

Marit Kapla, *Osebol: Voices from a Swedish Village*

Malcolm Gaskill, *The Ruin of All Witches: Life and Death in the New World*

Mark Mazower, *The Greek Revolution: 1821 and the Making of Modern Europe*

Paul McCartney, *The Lyrics: 1956 to the Present*

Brendan Simms and Charlie Laderman, *Hitler's American Gamble: Pearl Harbor and the German March to Global War*